Lecture Notes in Computer Science   13830

Founding Editors

Gerhard Goos

Juris Hartmanis

Editorial Board Members

The series Lecture Notes in Computer Science (LNCS), including its subseries Lecture Notes in Artificial Intelligence (LNAI) and Lecture Notes in Bioinformatics (LNBI), has established itself as a medium for the publication of new developments in computer science and information technology research, teaching, and education.

LNCS enjoys close cooperation with the computer science R & D community, the series counts many renowned academics among its volume editors and paper authors, and collaborates with prestigious societies. Its mission is to serve this international community by providing an invaluable service, mainly focused on the publication of conference and workshop proceedings and postproceedings. LNCS commenced publication in 1973.

Shelly Sachdeva · Yutaka Watanobe ·
Subhash Bhalla
Editors

Big Data Analytics in Astronomy, Science, and Engineering

10th International Conference on Big Data Analytics, BDA 2022
Aizu, Japan, December 5–7, 2022
Proceedings

 Springer

Editors
Shelly Sachdeva ⓘD
National Institute of Technology Delhi
New Delhi, India

Yutaka Watanobe ⓘD
University of Aizu
Fukushima, Japan

Subhash Bhalla ⓘD
University of Aizu
Fukushima, Japan

ISSN 0302-9743   ISSN 1611-3349 (electronic)
Lecture Notes in Computer Science
ISBN 978-3-031-28349-9   ISBN 978-3-031-28350-5 (eBook)
https://doi.org/10.1007/978-3-031-28350-5

This Springer imprint is published by the registered company Springer Nature Switzerland AG
The registered company address is: Gewerbestrasse 11, 6330 Cham, Switzerland

Preface

The volume of data that is managed by computer systems continues to grow with time. It has increased manyfold in recent times. This is due to advances in networking technologies, storage systems, the adoption of mobile and cloud computing and the wide deployment of sensors for data collection. As a result, there are five attributes of data that pose new challenges- volume, variety, velocity, veracity and value. To make sense of emerging big data to support decision-making, the field of Big Data Analytics has emerged as a key research and study area for industry and other organizations. Numerous applications of big data analytics are found in several diverse fields such as e-commerce, finance, healthcare, education, e-governance, media and entertainment, security and surveillance, smart cities, telecommunications, agriculture, astronomy, and transportation.

Analysis of big data raises several challenges such as how to process extremely large volumes of data, process data in real-time, and deal with complex, uncertain, heterogeneous and streaming data, which are often stored in multiple remote storage systems. To address these challenges, new big data analysis solutions must be devised by drawing expertise from several fields such as big data processing, data mining, database systems, statistics, machine learning and artificial intelligence. There is also an important need to build data analysis systems for emerging applications such as vehicular networks, social media analysis, and time-domain astronomy and to facilitate the deployment of big data analysis techniques in Artificial Intelligence and related applications.

The tenth International Conference on Big Data Analytics (BDA) was held on December 5–7, 2022. It was held jointly in virtual conference mode at the University of Aizu, Japan, the Indian Institute of Technology, Delhi (IITD) and at the National Institute of Technology, Delhi (NITD), India. Its proceedings in book form includes 14 peer-reviewed research papers and contributions by keynote speakers, and invited speakers. This year's program covers a wide range of topics related to big data analytics on themes such as: big data analytics, networking, social media, search, information extraction, image processing and analysis, spatial, text, mobile and graph data analysis, machine learning, and healthcare.

It is expected that research papers, keynote speeches, and invited talks presented at the conference will encourage research on big data analytics, stimulate the development of innovative solutions and their adoption in industry.

The conference received 70 submissions. The Program Committee (PC) consisted of researchers from both academia and industry from many different countries. Each submission was reviewed by at least two, and at most by three Program Committee members, and was discussed by PC chairs before taking the decision. Based on the above review process, the Program Committee selected 14 full papers. The overall acceptance rate was about 20%.

We would like to extend our sincere thanks to the members of the Program Committee and external reviewers for their time, energy and expertise in providing support to BDA 2022.

Additionally, we would like to thank all the authors who considered BDA 2022 as the forum to publish their research contributions. The Steering Committee and the Organizing Committee deserve praise for the support they provided. A number of individuals contributed to the success of the conference. We thank Marcin Paprzycki, Srinath Srinivasa, Sanjay Chawla, Divyakant Agrawal, and Huzur Saran for their insightful suggestions. We also thank all the keynote speakers and invited speakers. We would like to thank the sponsoring organizations, including the National Institute of Technology, Delhi (NITD), the Indian Institute of Information Technology, Delhi (IITD), India, the University of Aizu, Japan, and the Department of Computer Science at NITD, as they deserve praise for the support they provided.

The conference received valuable support from the University of Aizu, IIT Delhi and NIT Delhi for hosting and organizing the conference. At the same time, thanks are also extended to the faculty, staff members and student volunteers of the Department of Computer Science at the University of Aizu, IIT Delhi and at NIT Delhi for their constant cooperation and support.

December 2022

Shelly Sachdeva
Yutaka Watanobe
Subhash Bhalla

Organization

Patrons

Ajay K. Sharma (Director)   NIT Delhi, India
Toshiaki Miyazaki (President)   University of Aizu, Japan

General Chairs

Sanjiva Prasad   IIT Delhi, India
Shelly Sachdeva   NIT Delhi, India

Steering Committee

Srinath Srinivasa   IIIT Bangalore, India
Huzur Saran (Chair)   IIT Delhi, India
Prem Kalra   IIT Delhi, India
H. V. Jagadish   University of Michigan, USA
Divyakant Agrawal   University of California at Santa Barbara, USA
Arun Agarwal   University of Hyderabad, India
Subhash Bhalla (Chair)   The University of Aizu, Japan
Nadia Berthouze   UCL, UK
Cyrus Shahabi   University of Southern California, USA

Program Committee Chairs

Shelly Sachdeva   NIT Delhi, India
Yutaka Watanobe   University of Aizu, Japan

Organizing Chair

Shelly Sachdeva   NIT Delhi, India

Publication Chair

Subhash Bhalla   University of Aizu, Japan

Tutorial Chair

Punam Bedi   Delhi University, India
Chandra Prakash   NIT Delhi, India
Rishav Singh   NIT Delhi, India

Publicity Chair

Rashmi Prabhakar Sarode   University of Aizu, Japan
Shivani Batra   KIET Group of Institutions, India

Program Committee

D. Agrawal   University of California, USA
F. Andres   National Institute of Informatics, Tokyo, Japan
Nadia Bianchi-Berthouze   University College London, UK
Paolo Bottoni   University of Rome, Italy
Pratul Dublish   Microsoft Research, USA
L. Capretz   Western University, Canada
M. Capretz   Western University, Canada
Richard Chbeir   Pau University, France
William I. Grosky   University of Michigan-Dearborn, Michigan,
                                     USA
Jens Herder   University of Applied Sciences, Fachhochschule
                                     Dusseldorf, Germany
Masahito Hirakawa   Shimane University, Japan
Qun Jin   Waseda University, Tokyo, Japan
Srinath Srinavasa   IIIT Bangalore, India
Akhil Kumar   Pennsylvania State University, USA
Rishav Singh   NIT Delhi, India
Jianhua Ma   Hosei University, Tokyo, Japan
Anurag Singh   NIT Delhi, India
K. Myszkowski   Max-Planck-Institut fuer Informatik, Germany
T. Nishida   Kyoto University, Japan
Manisha Bharti   NIT Delhi, India

Baljit Kaur	NIT Delhi, India
Prakash Srivastava	IIET Group of Institutions, India
Rahul Katarya	DTU, Delhi, India
Vivek Shrivastava	NIT, Delhi

Sponsoring Institutions

National Institute of Technology, Delhi, India
Indian Institute of Technology, Delhi, India
University of Aizu, Japan

Contents

Information Interchange of Web Data Resources

Business Analytics

Data Science: Systems

Ontology Augmented Data Lake System for Policy Support

Apurva Kulkarni(✉) ⓘ, Pooja Bassin ⓘ, Niharika Sri Parasa ⓘ,
Vinu E. Venugopal ⓘ, Srinath Srinivasa ⓘ, and Chandrashekar Ramanathan ⓘ

International Institute of Information Technology, 26/C, Electronics City Phase 1,
Bangalore, Karnataka, India
{apurva.kulkarni,pooja.bassin,niharikasri.parasa,
vinu.ev,sri,rc}@iiitb.ac.in

Abstract. Analytics of Big Data in the absence of an accompanying framework of metadata can be a quite daunting task. While it is true that statistical algorithms can do large-scale analyses on diverse data with little support from metadata, using such methods on widely dispersed, extremely diverse, and dynamic data may not necessarily produce trustworthy findings. One such task is identifying the impact of indicators for various Sustainable Development Goals (SDGs). One of the methods to analyze impact is by developing a Bayesian network for the policymaker to make informed decisions under uncertainty. It is of key interest to policy-makers worldwide to rely on such models to decide the new policies of a state or a country (https://sdgs.un.org/2030agenda). The accuracy of the models can be improved by considering enriched data – often done by incorporating pertinent data from multiple sources. However, due to the challenges associated with volume, variety, veracity, and the structure of the data, traditional data lake systems fall short of identifying information that is syntactically diverse yet semantically connected. In this paper, we propose a Data Lake (DL) framework that targets ingesting & processing of data like any traditional DL, and in addition, is capable of performing data retrieval for applications such as Policy Support Systems (where the selection of data greatly affect the output interpretations) by using ontologies as the intermediary. We discuss the proof of concept for the proposed system and the preliminary results (IIITB Data Lake project Website link: http://cads.iiitb.ac.in/wordpress/) based on the data collected from the agriculture department of the Government of Karnataka (GoK).

Keywords: Big data · Ontology · Document retrieval · Data lake · Data analyses · Policy support system · Bayesian network

This work is supported by Karnataka Innovation & Technology Society, Dept. of IT, BT and S&T, Govt. of Karnataka, India.

1 Introduction

The Big data era left us with myriad challenges along the dimensions of processing, management, and effective retrieval [18]. Traditional database systems are not well equipped to handle the new ascending characteristics of data, such as volume, variety, and velocity. Considering these characteristics, conventional ETL techniques used in data warehouse systems are not the best tools for big data [11]. The notion of a "data lake" came in light of the difficulties associated with big data [13]. A data lake architecture gives us the flexibility to store data at any volume, regardless of its variety, speed, or veracity, which facilitates the storage of heterogeneous data from various sources under a single roof for diverse downstream applications [19, 21].

A typical example of a big heterogeneous data collection is the silos of data maintained by any government authorities. These kinds of data are often considered to be largely raw; meaning, not follow any particular standards or structure, and often they are cumbersome to deal with. In what follows, we will discuss the characteristics of raw data in general and a contemporary downstream activity we are often asked to perform on such data. We consider the open data related to the agriculture domain available from the Karnataka Government website[1] as our example.

Characteristics of Raw Data. Managing huge volumes, various data formats, and numerous data sources in the age of big data is a difficult task that is made even more challenging in the case of raw data. In our observation, the following are some of the important characteristics of the data we get from the public sector: (a) the data comes from a variety of sources, and those sources use different data formats, (b) common data points are collected over time, causing them to re-appear with different time stamps. (For example, the total yield of crops in an area is calculated for each season like Kharif, and Rabi over different timestamps.) This necessitates the management of data in relation to time as well, (c) Data is gathered at different levels of granularity (country, state, village, etc.) and hence it is challenging to receive data at regular intervals, (d) Understanding the geographic intervention is crucial for comprehending the data. (For example, the total yield for Gadag and total yield for Devagiri; in this case, Gadag is a district and Devgiri is a village.), (e) data is abundant, but not all of it may be required at a time for a decision.

Contemporary Downstream Activity. Understanding the impact of policy is as important as enacting it. Having accurate data makes it possible to decide on a policy, understand its indicators, and analyze its effects [3]. For example, the article[2] summarises the organic farming experiment conducted in Sri Lanka. The article highlights that the common people of Sri Lanka were experiencing severe fuel and food shortages (due to skyrocketing inflation) until there is a shift in the policy adopted by Sri Lankan government. The government identified the

---

[1] https://www.indiastat.com/data/agriculture.
[2] https://foreignpolicy.com/2022/03/05/sri-lanka-organic-farming-crisis/.

root cause of this issue (as the exhaustive use of chemical fertilizers[3]), and they came up with a policy decision to forcibly switch Sri Lanka to organic farming and ban all imports of chemical fertilizers. The experiment with organic farming resulted in a 40% decrease in rice yields and an equally significant decline in the yields of many other crops, which has exacerbated food shortages and aided in the economic collapse of the country.

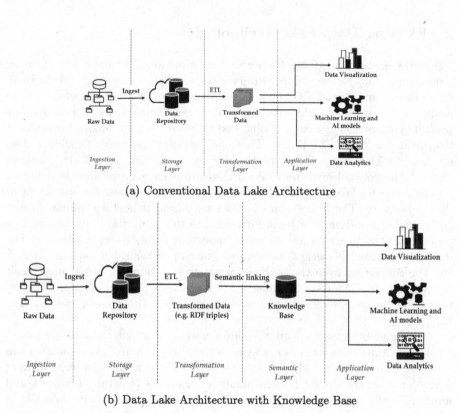

(a) Conventional Data Lake Architecture

(b) Data Lake Architecture with Knowledge Base

Fig. 1. Reference architecture.

The aforementioned downstream activity can be performed in a time-bounded manner (i) if the data required for the study is identified already, and (ii) if the policy enunciation method is already determined. However, under a generic circumstance, the data lake should be equipped to support features for the effective retrieval of data based on the specific demands of the downstream applications.

In this paper, we first discuss how the conventional data lake architecture had been evolved to support the requirements of modern applications. We will also look at the potential downsides of the data lakes while supporting such

---

[3] https://brownpoliticalreview.org/2022/11/sri-lankas-organic-farming-failure/.

applications. We then propose a new architecture for the data lake that can easily incorporate modern applications like the Policy Support System (PSS) (discussed in [3]). We observed that the data retrieval module of the data lake needs to be adapted inherently to data retrieval needs of the application. Our new architecture explored the use of description logic-based ontologies for solving such requirements, and hence the name "ontology-augmented data lake".

2   Existing Data Lake Technologies

The data lake technology has been widely adopted around the world to provide a data repository for open data. Projects such as data.europa.eu[4], World Bank Open Data[5], and the OECD Open Government Data project[6] collect data from countries around the world with the goal of developing a set of policies that promote transparency, accountability, and value creation by making government data available to all [14,17,20]. The most popular choice for providing data services was Data Lake. The typical structure of the data lake is showcased in Fig. 1a. The conventional data lake architecture is organized into four layers. At the ingestion layer, raw data is cleaned and ingested into the storage layer data repository. The stored data is then transformed into a common format (relational or non-relational) and forwarded to the application layer for further processing. The following are the most important points to emphasize: (a) The data is duplicated because both the original and transformed data are stored, (b) The conversion from the original format to the common format may result in data loss, (c) The data documents are transformed into a standard format. Given the massive size, the transformation could be a time/resource-consuming task.

Governments in various countries are curating their own data lakes through Open Data Initiatives such as data.gov in the United States, open.canada.ca in Canada, data.gov.in in India, and many more. These portals are critical from a data analysis standpoint because many governments (federal, provincial, and municipal) rely on them for policy making [1,6,9,16]. A traditional data lake is not enough when it comes to e-governance and AI modeling. It's crucial to comprehend how various data points relate to one another. The data lake architecture has been advanced to support the knowledge base in order to understand the semantic relationships between the data [2,4,10,22]. Figure 1b shows the general architecture of a knowledge base supporting a data lake. The raw data is transformed into comma-separated values with reference to the architecture shown in Fig. 1b, and then represented in triple form. The triples produce a data graph that depicts the semantic connections in the data. The application layer can query data and use it for additional analysis. In other words, the raw data in its pre-processing stage gets transformed into a knowledge graph. The end user then could make use of any semantic resolution approach to retrieve the most

---

[4] https://data.europa.eu/en.

[5] https://data.worldbank.org/.

[6] https://data.oecd.org/.

appropriate data. Albeit this approach has advantages in terms of being able to perform semantic-based information resolution, we may have to limit ourselves to such approaches alone. This will be more evident while considering that the data in its raw form (for instance, text documents) are the most comprehended form of information, and performing information extraction of such raw data may result in the loss of information.

Another reason to rethink the architecture of existing data lakes is the recent advancements in modern data processing systems [7,24–27]. The notion of unification in the modern big data system has resulted in developing a single entry point (tool or API) for different operations in a pipeline that were performed previously on multiple frameworks [27].

New Requirements. Following our understanding of government data, our research focused on developing a framework to assist policymakers in locating relevant documents from various data sources and developing their plans. Additionally, the framework also provides a no-code environment to analyze data from a variety of sources [8]. The organization of the research is as follows. Understanding the state of the art at the moment is the main goal of the following section. The suggested system and early findings are covered in the following section. The research work is concluded by discussing its potential future directions in the last section. Based on the base architectures depicted in Fig. 2, it is observed that data lakes handle data documents as data points. Without understanding the need, the entire data set is transformed into a common format. The data conversion may result in data loss and semantic loss. Instead of semantic mapping on data points, semantic mapping on documents would aid in the retrieval of the most relevant documents in their original form. Furthermore, if policymakers or data analysts deem these documents necessary for further processing, ETL operations might be done at this level. Because the data is in its original format, there is no transformed data or metadata to retain, resulting in minimal resource consumption. Pushing the ETL process after document retrieval can let analysts or policy maker focus on only the most important data points.

3 Proposed System

The proposed framework is split into two primary subsystems. The first subsystem is the Ontology-based Data Lake (DL), which aims to ingest, process, and retrieve the most relevant information. By creating pipelines and models on the retrieved documents, the second sub-system provides a no-code environment for Policy Enunciation and data analysis.

Figure 2 describes the framework for an ontology-augmented data lake system to support a policy support system. Primarily, there are three components in the figure: Data Lake, Exploratory Analytics, and Policy Enunciation. Component-A constitutes the ontology-augmented data lake part of the framework. Components B and C denote the PSS application integrated into the data lake architecture.

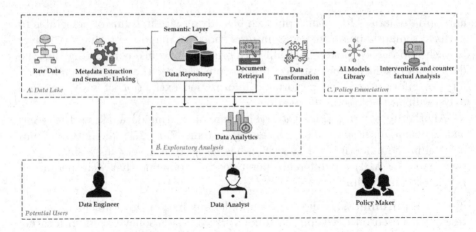

Fig. 2. Architecture of the proposed system.

3.1 Data Lake

The data lake lays the groundwork for semantic document retrieval. There are two main functionalities, namely Document Ingestion(DI) and Document Retrieval (DR). The domain knowledge is achieved from ontology. The concepts, relationships among concepts, and their description of all this semantic information are captured in the ontology. The ontology is used to achieve a semantic index for the retrieval of pertinent documents.

The ingestion includes structured (or unstructured) documents. The raw data is cleaned and then used for meta-data generation. Extracted metadata of the document is mapped to a generic set of latent concepts. These latent concepts are identified by clustering 10K Wikipedia articles. The top keywords of clusters define the latent concept. The metadata of the document and the latent concepts are compared to achieve documents-to-cluster mapping. The mappings are performed using the Wu-Palmer similarity metric. These clusters would act as the prime document classifier in the ingestion phase. The clusters remain static anchors to map various documents to the latent concept. Similarly, using the Wu-Palmer similarity metric, ontology concepts are mapped to latent concepts. Figure 3 elaborates the documents to cluster mapping and ontology to cluster mapping.

In the DR phase, the query string will be first mapped to the specific concepts in the ontology; this is with the assumption that each concept has an associated bag of words onto which semantic closeness could be checked. Once the search string identifies the concept, concept-to-cluster mapping is used to fetch relevant clusters. The cluster identifies all the documents using document-cluster-mapping.

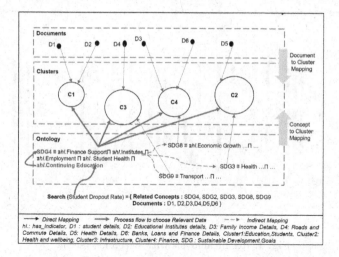

Fig. 3. Semantic linking of documents. (Color figure online)

In Fig. 3, the control flow to acquire documents related to "student dropout rate" is highlighted in blue. This input string is strongly related to the concept `Continuing_Education` which is an indicator of (hasIndicator^{-1}) SDG4. SDG4, as per its ontology definition, is related to topics such as students' health, educational information, scholarships, loans, and so on. Therefore, the OMR already maintains mappings from SDG4 to clusters $C1$, $C2$, $C3$, and $C4$. Hence, when there is a query that is related to SDG4, the documents linked with the associated clusters will be retrieved. In addition, the red arrows in the figure denote implicit relationships between the SDGs that were established by considering the latent topics shared by these concepts at the cluster level. This implicit relationship further helps in retrieving the pertinent data from multiple departments.

Our initial results reported in Fig. 4 are based on the agricultural data available from GoK. The dataset has 84 files totalling about 100MB in size, including 73 csv files, 6 pdf documents, and 5 SQL files. The ontology used (containing 115 concepts, 6 object properties, and approximately 27000 instances) in our study was manually curated by ontologists based on the SDG definitions available on the indiastat website[7], GoK. We consider Apache Lucene 4, an open-source IR tool [5], as the baseline for comparing our results. Our study here is limited to a selected set of keywords – that are not explicitly present in the ingested documents – from the agricultural domain. It is observed that the proposed ontology-mediated approach consistently gives us better output in terms of precision, recall, and F1 score when compared with the baseline approach.

---

[7] https://www.indiastat.com/.

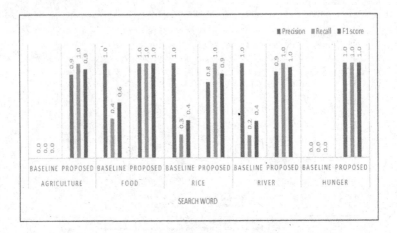

Fig. 4. Result analysis with baseline system.

3.2 Exploratory Data Analytics

Exploratory Data Analytics acts as a visualization subsystem. The data lake is utilized by data analysts to query, join, and merge datasets from various departments. This subsystem imparts a broad approach to identifying and solving problems. The subsystem has the ability to perform real-time analysis using AI/ML models and is primarily in charge of carrying out and providing initial investigations, observations, and correlations. Using visual representations like tabular views, graphs, charts, and maps, the exploratory data analytics subsystem accelerates the deep understanding of data by revealing trends and patterns, spotting anomalies, and testing hypotheses. With the help of the data, this subsystem also makes it possible to pinpoint areas that require attention or improvement and provides transparency regarding the variables that affect behavioral patterns.

Figure 5 depicts a comparative analysis of the prevalence of anemia and underweight among adolescent girls across regions of Karnataka. Figure 5a shows the percentage of adolescent girls with anemia across regions of Karnataka. The color intensity depicts the severity of anemia - the higher the intensity of the red color, the higher the percentage of anemic adolescent girls. Similarly, Fig. 5b shows the percentage of underweight adolescent girls across regions of Karnataka. It can be observed that being underweight among adolescents is mostly prevalent in the northern regions of Karnataka.

Another example of Exploratory Data Analytics is a story Fig. 6, that shows a series of figures with the percentage change in yield among villages over consecutive years i.e., 2017–22 in the Gadag district based on crop wise, agro-climatic zone wise, season wise, and irrigated type.

Figure 7 depicts an interactive, dashboard allowing users to understand how much quantity (Kgs in Metric Tonnes) of fertilizer (NPK) had been used for agriculture in districts of Karnataka from 2014 to 2019.

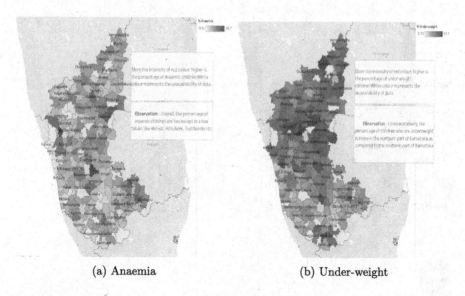

(a) Anaemia   (b) Under-weight

Fig. 5. Percentage of Anaemic and Under-weight Adolescent girls in Karnataka.

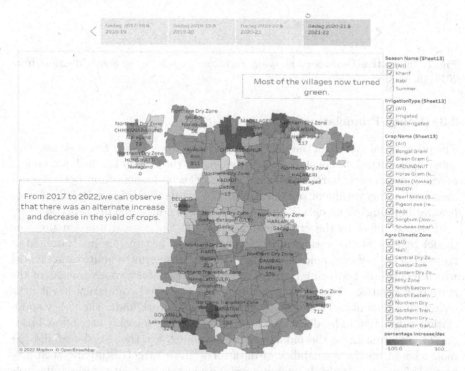

Fig. 6. Story depicting percentage change in yield among villages over 2017–2022 in Gadag district.

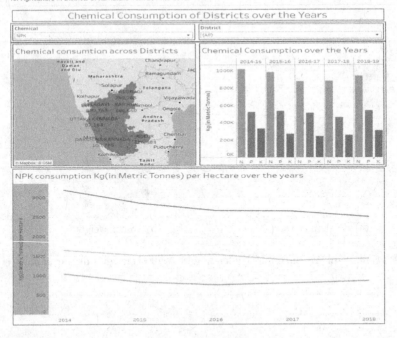

Fig. 7. Interactive Dashboard showing Fertilizer Consumption across districts from 2014–19.

3.3   Policy Enunciation

The primary subsystem of the Policy Support System is *Policy Enunciation*. This subsystem delves into constructing AI models for the end users which are policy-makers, researchers, bureaucrats, and other decision-makers from both government and non-government organizations. The availability of a well-defined data lake for building AI models is instrumental in the policy-making process.

Figure 8 refers to the crop yield model from the agriculture domain. The model represents the various factors that may be responsible for the yield of crops for a given geographic entity. Here, we are considering a multi-target model which focuses on two primary crops- wheat and rice, for various talukas of the state of Karnataka. Test the current status of yield and comparing it with the output of interventions is one of the main scopes of work under the policy enunciation subsystem. The data lake helps in retrieving relevant attributes based on the requirement of the model. These attributes are then used to build the model based on the availability of data for a given target. Studying such networks belonging to multiple domains and the generation of data stories influences policymakers significantly to take informed decisions. Similar to Fig. 8, we plan

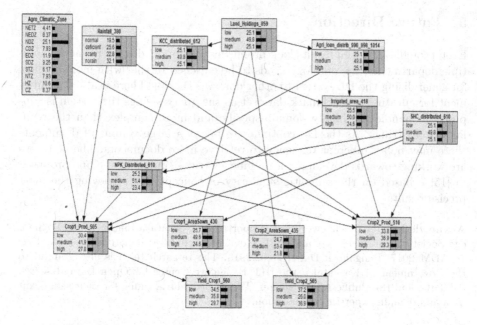

Fig. 8. Agriculture Crop Yield Bayesian network.

to build models for multiple domains such as health, education, etc. to study the downstream and lateral impacts within and outside the domains. The availability of data belonging to multiple domains as part of the data lake becomes advantageous in developing such use cases.

4   Conclusions

The study proposes an ontology-augmented data lake to help policymakers and data analysts establish a long-term policy support system. The proposed framework is implemented on data acquired from the Karnataka government. Using semantic linking, the user may easily access relevant documents in a unified manner, allowing them to explore and analyze pertinent information without having to access distinct data sources separately. The system provides a user interface with geospatial maps that allows the user to do interactive exploratory analytics across many data sources. The no-code environments allow policymakers and data analysts to access and analyze the most important data points for decision-making.

5 Future Direction

Even though our initial results look promising to support documents across multiple departments, the ontology needs to be enriched further with more axioms, for generalizing the DL system to other domains [12,23]. There is also a requirement for curating a benchmark data set as the query strings the system is supposed to handle are highly domain-specific and more complex than the regular search strings. As the DL continues to ingest a large volume of documents there may be an inherent tendency to retrieve more documents. Therefore, we are working towards developing a ranking metric – similar to the one proposed in [15] – based on the semantic similarity of concepts for retrieving the most precise results.

Acknowledgement. This work was supported by Karnataka Innovation & Technology Society, Dept. of IT, BT and S&T, Govt. of Karnataka, India, vide GO No. ITD 76 ADM 2017, Bengaluru; Dated 28.02.2018. The research team is also grateful to the Government of Karnataka, the IIIT Bangalore Center for Open Data Research (CODR), and the Public Affairs Center (PAC), Bengaluru, India, for their significant data and domain expertise collaboration.

References

1. Ali, A., Manzoor, D., Alouraini, A.: The implementation of government cloud for the services under e-governance in the KSA. Sci. Int. **33**(3), 249–257 (2021)
2. Bagozi, A., Bianchini, D., De Antonellis, V., Garda, M., Melchiori, M.: Personalised exploration graphs on semantic data lakes. In: Panetto, H., Debruyne, C., Hepp, M., Lewis, D., Ardagna, C.A., Meersman, R. (eds.) OTM 2019. LNCS, vol. 11877, pp. 22–39. Springer, Cham (2019). https://doi.org/10.1007/978-3-030-33246-4_2
3. Bassin, P., Parasa, N.S., Srinivasa, S., Mandyam, S.: Big data management for policy support in sustainable development. In: Sachdeva, S., Watanobe, Y., Bhalla, S. (eds.) International Conference on Big Data Analytics, pp. 3–15. Springer, Cham (2021). https://doi.org/10.1007/978-3-030-96600-3_1
4. Beheshti, A., Benatallah, B., Nouri, R., Tabebordbar, A.: CoreKG: a knowledge lake service. Proc. VLDB Endow. **11**(12), 1942–1945 (2018)
5. Bialecki, A., Muir, R., Ingersoll, G.: Apache lucene 4. In: Trotman, A., Clarke, C.L.A., Ounis, I., Culpepper, J.S., Cartright, M., Geva, S. (eds.) Proceedings of the SIGIR 2012 Workshop on Open Source Information Retrieval, OSIR@SIGIR 2012, Portland, Oregon, USA, 16 August 2012, pp. 17–24. University of Otago, Dunedin, New Zealand (2012)
6. Boldyreva, E., Gorbunova, N., Grigoreva, T.Y., Ovchinnikova, E.: E-government implementation in Spain, France and Russia: efficiency and trust level. In: SHS Web of Conferences, vol. 62, p. 11005. EDP Sciences (2019)
7. Carbone, P., Ewen, S., Fóra, G., Haridi, S., Richter, S., Tzoumas, K.: State management in apache Flink® consistent stateful distributed stream processing. PVLDB **10**(12), 1718–1729 (2017)

8. Daly, H.E.: Beyond Growth: The Economics of Sustainable Development. Beacon Press, Boston (2014)
9. De Blasio, E., Selva, D.: Implementing open government: a qualitative comparative analysis of digital platforms in France, Italy and united kingdom. Qual. Quant. **53**(2), 871–896 (2019)
10. Dibowski, H., Schmid, S.: Using knowledge graphs to manage a data lake. In: GI-Jahrestagung, pp. 41–50 (2020)
11. Diouf, P.S., Boly, A., Ndiaye, S.: Variety of data in the ETL processes in the cloud: state of the art. In: 2018 IEEE International Conference on Innovative Research and Development (ICIRD), pp. 1–5. IEEE (2018)
12. Fathy, N., Gad, W., Badr, N.: A unified access to heterogeneous big data through ontology-based semantic integration. In: 2019 Ninth International Conference on Intelligent Computing and Information Systems (ICICIS), pp. 387–392. IEEE (2019)
13. Giebler, C., Gröger, C., Hoos, E., Schwarz, H., Mitschang, B.: Leveraging the data lake: current state and challenges. In: Ordonez, C., Song, I.-Y., Anderst-Kotsis, G., Tjoa, A.M., Khalil, I. (eds.) DaWaK 2019. LNCS, vol. 11708, pp. 179–188. Springer, Cham (2019). https://doi.org/10.1007/978-3-030-27520-4_13
14. Idowu, L.L., Ali, I.I., Abdullahi, U.G.: A model and architecture for building a sustainable national open government data (OGD) portal. In: Proceedings of the 11th International Conference on Theory and Practice of Electronic Governance, pp. 352–362 (2018)
15. Kumar, R., Sharma, S.C.: Smart information retrieval using query transformation based on ontology and semantic-association. IJACSA **13**(4), 388 (2022)
16. Majeed, B., Niazi, H.A.K., Sabahat, N.: E-government in developed and developing countries: a systematic literature review. In: 2019 International Conference on Computing, Electronics & Communications Engineering (iCCECE), pp. 112–117. IEEE (2019)
17. Miller, R.J.: Open data integration. Proc. VLDB Endow. **11**(12), 2130–2139 (2018)
18. Mureddu, F., Schmeling, J., Kanellou, E.: Research challenges for the use of big data in policy-making. Transform. Gov. People Process Policy **14**(4), 593–604 (2020)
19. Nargesian, F., Zhu, E., Miller, R.J., Pu, K.Q., Arocena, P.C.: Data lake management: challenges and opportunities. Proc. VLDB Endow. **12**(12), 1986–1989 (2019)
20. Peña-López, I., et al.: Open, useful and re-usable data (ourdata) index: 2019 (2020)
21. Sawadogo, P., Darmont, J.: On data lake architectures and metadata management. J. Intell. Inf. Syst. **56**(1), 97–120 (2021)
22. Schmid, S., Henson, C., Tran, T.: Using knowledge graphs to search an enterprise data lake. In: Hitzler, P., et al. (eds.) ESWC 2019. LNCS, vol. 11762, pp. 262–266. Springer, Cham (2019). https://doi.org/10.1007/978-3-030-32327-1_46
23. Thirumahal, R., Sudha Sadasivam, G., Shruti, P.: Semantic integration of heterogeneous data sources using ontology-based domain knowledge modeling for early detection of COVID-19. SN Comput. Sci. **3**(6), 1–13 (2022)
24. Venugopal, V.E., Theobald, M., Chaychi, S., Tawakuli, A.: AIR: a light-weight yet high-performance dataflow engine based on asynchronous iterative routing. In: 32nd IEEE SBAC-PAD, Portugal, 9–11 September 2020, pp. 51–58. IEEE (2020)

25. Venugopal, V.E., Theobald, M., Tassetti, D., Chaychi, S., Tawakuli, A.: Targeting a light-weight and multi-channel approach for distributed stream processing. J. Parallel Distributed Comput. **167**, 77–96 (2022)
26. White, T.: Hadoop: The Definitive Guide, 1st edn. O'Reilly Media Inc., Sebastopol (2009)
27. Zaharia, M., et al.: Resilient distributed datasets: a fault-tolerant abstraction for in-memory cluster computing. In: Proceedings of the 9th USENIX Conference on Networked Systems Design and Implementation, NSDI 2012 (2012)

Explorations in Active Learning Applied to Image Classification

Adriana Klimczak[1], Marcel Wenka[1], Maria Ganzha[1,2] (iD),
and Marcin Paprzycki[2,3(✉)] (iD)

[1] Department of Mathematics and Information Science, Warsaw University
of Technology, Warsaw, Poland
`maria.ganzha@pw.edu.pl`
[2] Systems Research Institute Polish Academy of Sciences, Warsaw, Poland
`marcin.paprzycki@ibspan.waw.pl`
[3] Warsaw Management University, Warsaw, Poland

Abstract. While development of very large models is the core of today's artificial intelligence, very often the cost of model training is being raised. In this context, active learning is pointed to as a method to maximize model quality, while minimizing the amount of resources needed to train it. The aim of this contribution is to systematically compare performance of active learning applied to the image classification task for three datasets.

Keywords: Frugal AI · Neural networks · Active learning · Image classification

1 Introduction

Today, the core direction of research in, broadly understood, artificial intelligence (AI) involves extremely large models, which are trained on ever increasing datasets (see, for instance, GPT-3 [7]). In particular, such models materialize in the context of image and natural language processing. However, it is important to realize the "costs" of gathering the data, labeling it, persisting, and training the model(s). This is why only few, very large, companies have resources that allow them to train models with billions of parameters. Moreover, it should be noted that the required energy consumption has detrimental effect on the environment.

Separately, it is easy to realize that (1) humans do not learn "image classification" by associating millions of labels with petabytes of images, and (2) the amount of available data is systematically increasing, bringing the question of long-term usability of already trained models. Obviously, one can stipulate that models can be up-trained, but this brings back the question of the cost of labeling of the "incoming images".

This work was funded in part by the Centre for Priority Research Area Artificial Intelligence and Robotics of Warsaw University of Technology within the Excellence Initiative: Research University (IDUB) programme.

In this context, *active learning* has been considered. Active learning belongs to the category of semi-supervised learning, where only limited number of instances is actually labeled [10]. Specifically, during the model training, an *oracle* is asked to label selected samples. This results in slow increase of the size of the training dataset, to the point where the model provides an acceptable accuracy. In this context, the main contribution of this work is to present results of systematic experimentation with active learning applied to three image datasets.

Here, in Sect. 2, active learning is introduced and pertinent literature cited. Next, datasets, their preprocessing, and the experimental setup are summarized (in Sect. 3). This is followed, in Sect. 4, by presentation of results. Finally, Sect. 5, summarizes key findings and concludes the paper.

2 Active Learning – Introduction and Literature Review

Regardless of the specific model that is trained, the two main aspects of active learning are: (1) how to select samples to be added to the training set, and (2) what stopping criterion should be used. Let us start from sample selection. It has been stipulated the data elements to be labeled should be prioritized on the basis of them being "most informative", where "informativeness" can be understood as gain in model accuracy, after adding given element to the training set. Here, various approaches to prioritizing data elements to be labeled have been proposed. In particular, they aim at selecting samples near boundaries between classes [6].

The metrics used to select samples are called *uncertainty measures* [11]. In what follows, results of experiments with two very popular uncertainty measures are reported. First of them is the *largest margin uncertainty* (LMU), which is presented in Eq. 1.

$$\phi_{LM}(x) = P_\theta(y_1^*|x) - P_\theta(y_{min}|x) \qquad (1)$$

Here, the assumption is that if the probability of membership of the sample element in the most possible class is much larger than the probability of membership in the least possible class, then the model is confident about the sample's class membership [6]. On the other hand, if the probability of membership in the most possible class is not substantially larger than the probability of membership in the least possible class, then the model is not confident about class membership of the sample. Hence, the sample with the lowest LMU value is to be selected and sent to the oracle to be labeled.

The second measure is the *least confidence uncertainty* (LCU), which is represented in Eq. 2. This measure considers only at the most possible class and targets the samples with lowest probability of being assigned to that class.

$$\phi_{LC}(x) = 1 - P_\theta(y_1^*|x) \qquad (2)$$

In addition to the two uncertainty measure-based approaches (MLU and LCU), random sample selection has been experimented with. The latter can be treated as the generic baseline.

As noted, labeling of the selected samples is performed by the *oracle*, which may be an expert or an automated service. Here, the oracle is a service, which returns the correct label (based on the labels available in the original dataset).

The second issue in active learning is the stopping criterion. Obviously, in an ideal situation, the oracle would be "engaged" with the model, by being asked to label samples. Hence, the oracle itself could be tasked with deciding when to stop the training. However, such approach does not scale to real-life deployments. Therefore, other approaches have been proposed.

The basic approach [8], uses a predetermined budget, representing the number of queries that can be submitted. This approach is flawed in two ways. First, there may already be enough data to train the model and asking for more (to use the allotted budget) may result in oversampling [3]. This is particularly bad, since the goal of active learning is to limit the number of instances that have to be labeled. Second, after spending the available budget, the model may not achieve satisfying accuracy.

To remedy these problems, instead of predefining the number of instances to be queried, it is possible to specify the cost of querying the oracle and to calculate the (diminishing) error of prediction with further queries [3]. In this way it is possible to determine when the model should stop training.

However, due to the limitations in the size of the contribution, the effects of stopping criterion have not been reported. Nevertheless, they have to be kept in mind when real-life active learning deployments are to be deployed.

3 Datasets and Experimental Setup

In the experiments, three datasets have been used: CIFAR-10, CIFAR-100 and Caltech-256. The simplest one of them is *CIFAR-10*, which is a subset of a dataset of 80 million images [4]. CIFAR-10 contains 60,000 color images, equally distributed into 10 (mutually exclusive) classes – 6,000 images per class. All images are of the same size – 32×32 pixels. The dataset is divided into five training batches and one test batch, each with 10,000 images. The test batch consist of 1000 images randomly-selected from each class. The training batches contain the remaining images in random order, while some training batches may contain more images from one class than the another.

The *CIFAR-100 dataset* is an intermediate difficulty dataset. It is the subset of the same 80 million images. It also contains 60,000, but distributed into 100 classes (600 images per class). The images are also 32×32 pixels. Similarly to CIFAR-10, the dataset is split into five training batches and one test batch, where the test batch was randomly selected from the training set.

The *Caltech-256 dataset* is the most complex and consists of 30,607 color images, representing 256 classes [2]. In contrast to the CIFAR datasets, the classes are not equal in size. However, there are at least 80 images per class.

3.1 Preprocessing

In the case of both CIFAR datasets no preprocessing was needed. The images were loaded and transformed into $32 \times 32 \times 3$ files. The last dimension corresponded to the values of red, green and blue channels.

Since images in Caltech-256 differ in size, ithey have been unified. To avoid distorting them too much, the following transformations were applied: 20% zoom range (in or out), 20% width and height shift, horizontal flip (but not vertical) and 20° rotation. The images were then resized to 192×192 pixels so as not to lose too much detail (the images were much bigger on average).

3.2 Model Architecture

The model architecture was inspired by the state-of-the-art networks Alexnet [5]. Specifically, 2D Convolutional Layers (CL) were paired with Batch Normalisation (BN). Max Pooling (MP) and Dropout (D) were applied after every pair. The used model was composed of 3 repetitions of: CL+BN+CL+BN+MP+D. The number of filters increased with each CL+BN pair (i.e. 32, 64 and 128). The dropout rate also increased (i.e. 0.2, 0.3 and 0.4). Each convolutional layer used Rectified Linear Unit activation function, kernel size of 3×3, and 2×2 pool size in MP layers. The 3 pairs of convolutions were followed by: Flatten, Dense (128 neurons, ReLU activation), BN, D (0.5 rate), Dense (10 neurons), and Adam optimiser (0.001 learning rate, with categorical cross entropy). Experiments were performed on Google Colab Pro.

Note that the model is "relatively small" (compared to, for instance, Inception V3 or VGG-16). However, reaching the highest possible accuracy was not the goal of this work. Therefore, it was assumed that model is acceptable if, for each dataset, it could predict correct classes for at least 50% observations (for all data elements being labeled).

4 Experimental Results

The overarching goal of performed experiments was to systematically study, for the CIFAR-10, CIFAR-100 and Caltech-256 datasets, performance of active learning. In this context, the following tests have been performed:

1. comparison of accuracy for different (systematically increasing) number of labeled examples (ranging from a very small subset, up to the whole dataset);
2. analysis of effects of the two uncertainty measures: LCU and LMU used for sample selection;
3. comparison of accuracy achieved with labeling of a random batch data elements vs. accuracy achieved when applying active learning-based incremental labeling (until the same number of elements is labeled);
4. study of accuracy achieved when the active learning trained model is applied to the remaining data.

4.1   Experiments with CIFAR-10 Dataset

In the experiments, in each iteration, 500 samples has been selected, labeled and added to the training set. This was followed by model retraining. Both uncertainty measures were tested: LCU (Eq. 2) and LMU (Eq. 1), as well as random sampling. Results for CIFAR-10 have been presented in Fig. 1.

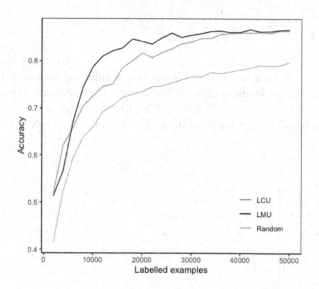

Fig. 1. Results of testing different metrics on CIFAR-10.

As expected, for both uncertainty measures, the best accuracy was achieved when the whole CIFAR-10 dataset has been labeled and model training turned into supervised one (baseline case). However, use of either uncertainty metrics delivered very good results with only half of the dataset labeled. Here, for 25,000 labeled images, for the LMU, 85.99% accuracy was reached, while for the LCU, the accuracy was 82.62%. Hence, use of LMU resulted in accuracy 0.99% points lower than the baseline. Finally, it can be clearly seen that random labeling was much worse than using either uncertainty measure.

Results for specific breakpoints are reported in Table 1. Here, for 5% (2,500 labeled samples) the results were low – 51.87% and 51.30%, for the LMU and the LCU, respectively. However, for 25% of the original dataset (12,500 labeled samples) the accuracy has reached 81.03% for the LMU, and 74.58% for the LCU. In comparison, William H. Beluch et al. [1] reported accuracy of 85% with 10,500 images (21% of the dataset). However, there, a much more robust architecture. Another work [9] reported accuracy of 77% with 25% of the dataset labeled, which is 4.03pp lower than the results obtained for the LMU-based approach.

Table 1. CIFAR-10 comparison between LMU and LCU.

% of the whole dataset	LMU [%]	LCU [%]
5	51.87	51.30
10	67.02	65.82
25	81.03	74.58
50	85.99	82.62
75	86.15	85.95

Based on these results, an experiment was conducted in which 12,500, randomly selected data elements (25% of the dataset) have been labeled "at once". This experiment was to provide additional information how active labeling differs from random labeling of a batch of data. The results are presented in Fig. 2.

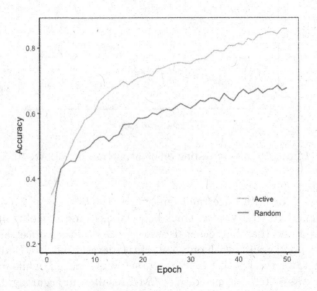

Fig. 2. Comparison of random vs. active LMU-based labeling on CIFAR-10.

The final accuracy of the model with random batch labeling was equal to 67.78%. It can be clearly seen that the accuracy achieved with LMU-based labeling was much higher. More detailed results are summarized in Table 2.

Here, the best accuracy, equal to 67.78%, is 19.2% points worse than baseline, and 13.25pp worse than in the case of LMU-based active approach. These results show that, for CIFAR-10, active learning leads to better results than random labeling, even when choosing data in one batch rather than iteratively. Overall, results presented here are similar to the ones reported in [9].

Table 2. CIFAR-10 training on randomly chosen 12500 labeled data.

Metric	Random	Active
Test accuracy	67.78%	81.03%
Train accuracy	70.60%	80.26%
Test loss	0.9101	0.6792
Train loss	0.6778	0.5520

Since after labeling 25% of data, accuracy better than 80% has been achieved, the question if using this model to label the remaining 75%, was explored. Here, active learning, with LMU uncertainty measure, used to select next batch of elements to be labeled, followed by model up-training, was applied. The results (accuracy and loss) have been presented in Fig. 3.

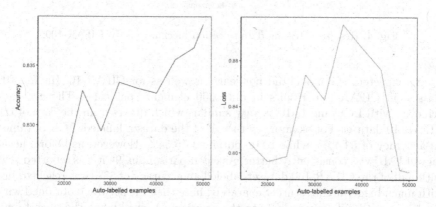

Fig. 3. CIFAR-10, LMU with auto-labeling the remaining part of the data.

Application of active learning trained model, and process, to label remaining data was relatively successful, as accuracy of 83.85% has been reached (improvement of 2.82% points). However, auto-labeling not always helps as it might happen, that in a specific batch, large number of samples would be improperly labeled, disrupting the model's performance (as it happened between 25,000 and 30,000 data elements).

4.2 Experiments with CIFAR-100 Dataset

After CIFAR-10, analogous experiments had been performed for the CIFAR-100 dataset. First, the comparison between LMU, LCU and random selection was explored. In each iteration additional 500 elements were selected, until the whole dataset was labeled. The results are presented in Fig. 4.

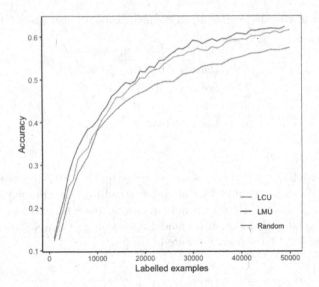

Fig. 4. Results of testing data selection mechanisms for CIFAR-100.

، As expected, the model did not learn as well as for CIFAR-10. Having 100 classes, in CIFAR-100, results in only 600 elements per class. The accuracy achieved with LCU and LMU is very similar, which differs from the case of the CIFAR-10 dataset. For example, with half of the dataset labeled, LCU obtained an accuracy of 54.40%, while LMU obtained 55.54%. However, as before, accuracy of LMU was consistently better. As a comparison, in [9], it was reported that with half of the CIFAR-100 dataset labeled an accuracy of 57% was reached, but with much bigger architecture. Separately, note that the accuracy obtained with random sampling is much closer to the accuracy reached when the uncertainty measures have been applied.

Additionally, the accuracy reached for the same breakpoints of percentage of labeled dataset, is reported in Table 3.

Table 3. CIFAR-100 comparison between LMU and LCU.

% of the whole dataset	LMU [%]	LCU [%]
5	21.87	19.99
10	31.46	26.51
25	46.07	43.90
50	55.54	54.40
75	59.20	60.05

With 25% of the dataset labeled, LMU achieved an accuracy of 46.07%, while LCU reached 43.90%. This, for LMU, is worse by 16.06% points from the baseline. However, the accuracy reached with half of the dataset labeled was 55.54% for LMU and 54.40% for LCU. This is close to the baseline (62.31%).

For CIFAR-100 the threshold of 25,000 labeled samples, was chosen for the test of random batched versus active learning based labeling. The results of the experiment are shown in Fig. 5 and Table 4.

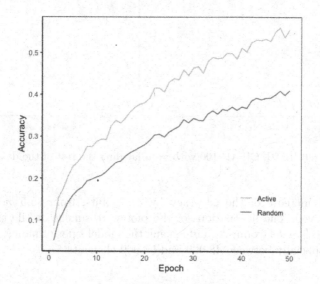

Fig. 5. Comparison of random vs active labeling on CIFAR-100.

Table 4. CIFAR-100 training on randomly chosen 25000 labeled data.

Metric	Random	Active
Test accuracy	37.79%	66.65%
Train accuracy	41.08%	55.54%
Test loss	2.3669	0.6665
Train loss	2.2538	0.5584

Here, similar conclusion can be reached. Active learning based training works better than random batching. The accuracy achieved with random labeling of 25,000 examples was equal to 41.08%, which is worse by 14.46% points from active-learning based and worse from the baseline by 21.23pp.

Finally, an experiment with active learning-based auto-labeling the remaining data has been performed. Here, the threshold of 50% of the whole dataset was chosen. The results are shown in Fig. 6.

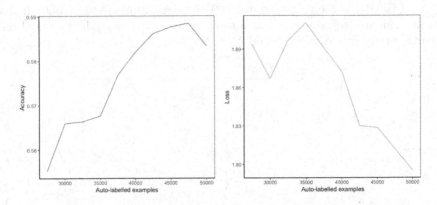

Fig. 6. LMU, CIFAR-100 with auto-labeling the rest of the data.

Here, an increase of the accuracy by 3% points, from 55.54% to 58.85% can be observed. The fluctuations of the plot were smaller, while the effect of wrongly auto-labeled examples, influencing the model's performance, can be also observed, especially between 48,000 and 50,000 elements.

4.3   Experiments with Caltech-256 Dataset

The last dataset was Caltech-256. The results obtained with application of the two uncertainty measures and with random sampling are represented in Fig. 7

As can be seen, this dataset is the most difficult to train the model. This was expected as Caltech-256 has 256 classes and the number of examples per class not only smaller, but laso unbalanced. Overall, with half the dataset (12,000 labeled elements) LMU reached accuracy of 41.44%, while LCU obtained 36.04%. The best (baseline) accuracy, achieved for the fully labeled dataset, was 52.25%.

As previously, random labeling proved to be the worst. Here, it can be conjectured that the performance differences may be related to the characteristics of the dataset, which had small unbalanced classes. Thus, it became particularly important to select elements that the model was "really not certain about" and label them first. This explains the bad performance of random sampling, and the fact that the best performance occurred for LMU-based sampling.

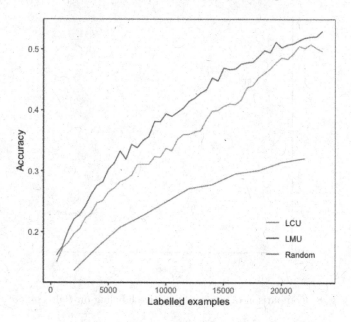

Fig. 7. Results of testing different metrics on Caltech-256.

Table 5. Caltech-256 comparison between LMU and LCU.

% of the whole dataset	LMU [%]	LCU [%]
5	20.15	18.20
10	22.82	20.47
25	33.28	28.20
50	41.44	36.04
75	49.47	46.79

Similarly to the previous dataset, performance at selected breakpoints has been summarized in Table 5.

Here, the accuracy achieved with 25% of data is not satisfactory; 33.28% for LMU and 28.20% for LCU. It is a significant decrease when compared to the baseline of 52.25%. However, with 75% of the dataset (18,364 labeled elements), an accuracy of 49.47% has been reached for the LMU-based approach. It is lower than the best accuracy by only 2.78% points.

For the sake of similarity, the threshold of half of the dataset being labeled, with an accuracy at 41.44% (for the LMU measure) became the starting point for the experiment comparing accuracy of active versus random batch labeling. The results have been summarized in Fig. 8.

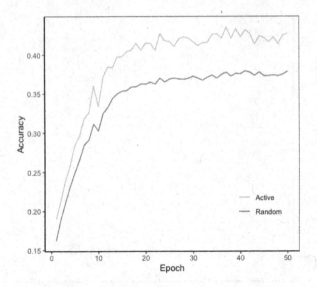

Fig. 8. Comparison of random vs active labeling on Caltech-256.

This experiment once more confirmed that iterative active learning is more effective than random batch labeling. However, it can also be seen that, as expected, the number of examples per class is very important, as the difference between batch random and active labeling is the smallest among all datasets.

Table 6 shows detailed results (accuracy and loss) of random and active learning-based labeling of 12,000 images for Caltech-256. The accuracy achieved that way was 37.88% which is 14.37% points worse than the baseline, and 3.56 worse than when active learning is used.

Table 6. Caltech-256 training on randomly chosen 12,000 labeled data.

Metric	Random	Active
Test accuracy	37.88%	41.44%
Train accuracy	90.14%	98.40%
Test loss	4.1924	5.4317
Train loss	0.3595	0.0550

The last experiment tested the auto-labeling of samples left after actively labeling half of the Caltech-256 dataset. The results can be seen in Fig. 9.

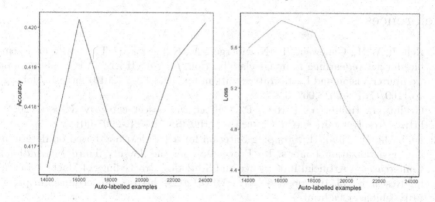

Fig. 9. LMU, Caltech-256 with auto-labeling the rest of the data.

Here, the results are quite peculiar. There is a deep fall in accuracy between 16,000 and 20,000 labeled elements. It shows clearly that auto-labeling the remaining part of data, even when selecting samples with an uncertainty measure, may not succeed. Caltech-256 is a complex dataset with small (unbalanced) number of examples per class. Hence, the model that had only around 47 labeled pictures per class "to work with" might often misclassify an image during automatic labeling. This, in turn, made the model significantly less accurate.

5   Concluding Remarks

The aim of this work was to systematically explore potential of active learning applied to the image classification. Experiments have been run using three datasets, with increasing level of difficulty to train the model, CIFAR-10, CIFAR-100 and CalTech-256. The main findings can be summarized as follows. (1) Active learning can allow successful training of models with limited amount of labeled data. (2) However, the quantity of labeled data that is needed to reach accuracy similar to the one that can be achieved using supervised learning, varies and depends on the characteristics of the dataset. (3) Order in which data is labeled matters. Elements that are added to the training set first, should be these for which the model has "most doubts" to which class they belong. Here, approach based on the LMU uncertainty measure produced slightly better results. At the same time, for datasets with small and/or unbalanced classes, it is important to select elements to be labeled in such a way to keep the training set balanced. (4) Using trained model to label the remaining data may improve model quality. However, results depend on the quality of the model and the properties of the dataset.

In the future, more research is needed how to select best element to be added to the training set. This should consider also characteristics of the dataset itself. Moreover, experiments with different model architecture may also be performed (for different datasets). Finally, the question how to apply principles of active learning to a steady stream of data, e.g. generated in IoT ecosystems should be explored.

References

1. Beluch, W.H., Genewein, T., Nurnberger, A., Kohler, J.M.: The power of ensembles for active learning in image classification. In: 2018 IEEE/CVF Conference on Computer Vision and Pattern Recognition, pp. 9368–9377 (2018). https://doi.org/10.1109/CVPR.2018.00976
2. Griffin, G., Holub, A., Perona, P.: Caltech-256 object category dataset (2007). https://resolver.caltech.edu/CaltechAUTHORS:CNS-TR-2007-001
3. Ishibashi, H., Hino, H.: Stopping criterion for active learning based on deterministic generalization bounds. In: Proceedings of the Twenty Third International Conference on Artificial Intelligence and Statistics. Proceedings of Machine Learning Research, vol. 108, pp. 386–397. PMLR (2020). https://proceedings.mlr.press/v108/ishibashi20a.html
4. Krizhevsky, A.: Learning multiple layers of features from tiny images (2009). https://www.cs.toronto.edu/~kriz/learning-features-2009-TR.pdf
5. Krizhevsky, A., Sutskever, I., Hinton, G.E.: Imagenet classification with deep convolutional neural networks. In: Pereira, F., Burges, C., Bottou, L., Weinberger, K. (eds.) Advances in Neural Information Processing Systems, vol. 25. Curran Associates, Inc. (2012). https://proceedings.neurips.cc/paper/2012/file/c399862d3b9d6b76c8436e924a68c45b-Paper.pdf
6. Miller, B., Linder, F., Walter R. Mebane, J.: Active learning approaches for labeling text: review and assessment of the performance of active learning approaches. https://websites.umich.edu/~wmebane/Paper_Active_Learning_Approaches_for_Labeling_Text.pdf
7. OpenAI: OpenAI API (2020). https://openai.com/blog/openai-api/
8. Schumann, R., Rehbein, I.: Active learning via membership query synthesis for semi-supervised sentence classification. In: Proceedings of the 23rd Conference on Computational Natural Language Learning (CoNLL), Hong Kong, China, pp. 472–481. Association for Computational Linguistics (2019). https://doi.org/10.18653/v1/K19-1044. https://aclanthology.org/K19-1044
9. Sener, O., Savarese, S.: Active learning for convolutional neural networks: a core-set approach. https://doi.org/10.48550/ARXIV.1708.00489. arXiv:1708.00489
10. Settles, B.: Active learning literature survey. Computer Sciences Technical Report 1648, University of Wisconsin-Madison (2009)
11. Sheng, V.S., Provost, F., Ipeirotis, P.: Get another label? Improving data quality and data mining using multiple noisy labeler. In: 14th ACM SIGKDD International Conference on Knowledge Discovery and Data Mining (KDD), pp. 614–662 (2008)

Discovery of Small Signals in Big Backgrounds

Milind V. Purohit[1,2](✉) ⓘ

[1] Okinawa Institute of Science and Technology, Tancha 1919-1, Onna-son, Kunigami-gun, Okinawa 904-0495, Japan
[2] Department of Physics and Astronomy, University of South Carolina, Columbia, SC 29208, USA
purohit@sc.edu

Abstract. Discovery of new phenomena typically consists of finding a small signal of data in processes which naturally produce similar data with much larger rates; these latter "events" we call background. The statistical problem, or data analytics problem, lies in discerning the signal in the large background i.e., the proverbial needle in a haystack. We see here how the analysis benefits from high dimensionality via successively reducing the background as we go higher in number of dimensions.

Keywords: Small signals · Big backgrounds · Discovery · Supersymmetry (SUSY) · High dimensionality

1 Introduction

As the illustrative setting for describing the concept we use data typically obtained in the particle physics experiments which started today's revolutionary big data acquisition. Indeed, these were the earliest large collections of data, certainly predating the enormous commercially driven datasets that we see today. In the 1960's particle physics experiments such as the one that discovered the Ω^- particle via interactions photographed in bubble chambers, [1] utilized the fact that the mass of the newly discovered particle was unique and typically higher than the mass of known particles. That analysis was helped by the accurate measurement of momenta via the verity of track positions and the strength of the magnetic field. This leads to unmistakably good data, and attendant backgrounds were reduced by the precision in the measurement of the particle mass. This mass is simply the square root of the 4-momenta summed and then squared:

$$m_{\Omega^-}^2 = (p_1 + p_2 + ...)^2 \tag{1}$$

where the 4-momenta p_i are of decay products.

Today, we seek "new physics" in the form of particles that are not described by the Standard Model (SM), [2] such as those that comprise supersymmetry (SUSY) [3]. In a typical version of SUSY particle production and decay, a massive

S. Sachdeva et al. (Eds.): BDA 2022, LNCS 13830, pp. 31–37, 2023.
https://doi.org/10.1007/978-3-031-28350-5_3

supersymmetric particle and its antiparticle are produced as a pair, and these subsequently decay until the lightest supersymmetric particle (the LSP) escapes the detector as a dark matter particle. If we imagine a chain of decays, then in the final step the LSP could be detected by missing energy and momentum and the chain could be reconstructed as we go further up if the remaining particles are charged or otherwise detected and the other LSP has low momentum that can be neglected.

Back in 1982, when I was still a graduate student, a magnetic monopole had reportedly just been observed [4]. Not long thereafter I heard of an experiment searching for a magnetic monopole which was also supersymmetric. Such was my hubris at the time that I remember chuckling at the thought of discovering not just monopoles, but also SUSY, all in one go. Either one was hard enough, so hey, why not go for both at the same time! Perhaps this paper will serve to convince skeptics that searching for more than we plan for is not such a bad idea after all.

In an ideal world, our data would have a large signal over an almost indiscernible background. For example, see Fig. 1 which is reminiscent of signals in a typical b-quark experiment [5]. On the other hand, the typical discovery situation is closer to the discovery signal for the Higgs boson [6]. An early version of the experimental data for the figure just referred to might look more like Fig. 2.

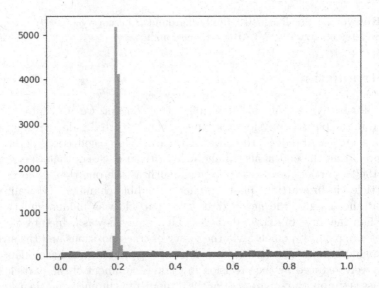

Fig. 1. A large Gaussian signal on a small background.

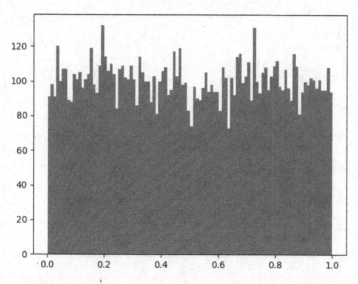

Fig. 2. A small Gaussian signal on a large flat background; the signal is at the same location as in the previous plot. The central question of discovery is how large and where is this "new" signal? Can we even be sure it exists?.

2   The Curse of Dimensionality

Wikipedia, the free encyclopedia, says that "The curse of dimensionality refers to various phenomena that arise when analyzing and organizing data that do not occur in low-dimensional settings such as the three-dimensional space of everyday experience. The expression was coined by Richard E. Bellmann when considering problems in dynamic programming." [7,8].

Backgrounds tend to be slowly changing, but complex in origin, at least let us assume so. In this case, we can imagine that we can transform/scale/shift the (mass) variable to get an approximately flat background in the range [0, 1].

We will see how to change the curse to a "Boon of Dimensionality".

Consider a small cluster of 5 "signal" events embedded in a uniformly distributed background of 100 events. These can be displayed in one, two, and three dimensions as Figs. 3, 4, and 5 respectively.

3   Criterion for Discovery

Imagine we have a unit hypercube with
N points (mostly background)

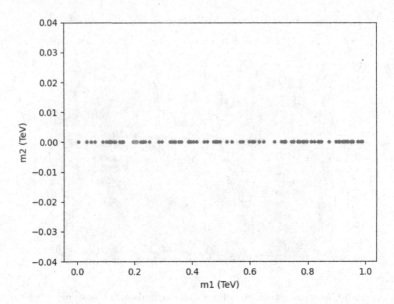

Fig. 3. A small cluster of 5 events on a flat background of 100 events in one dimension.

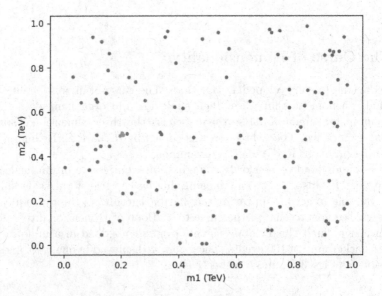

Fig. 4. A small cluster of 5 events on a flat background of 100 events in two dimensions.

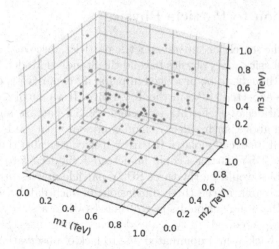

Fig. 5. A small cluster of 5 events on a flat background of 100 events in three dimensions.

D dimensions

v is the volume "containing" the signal. We assume that the signal is clustered near a point in all dimensions, and v is the volume that contains the bulk of the signal.

n_s is the expected number of signal events

The probability p of finding a background point in a given volume v equals v (for the unit hypercube). The probability of finding n_s points in the volume is thus v^{n_s}. If this is sufficiently small, the chance of finding at least one such volume is v^{n_s-1}. To emphasize the point about small probabilities, we can say that

$$v^{n_s-1} \ll 1 \tag{2}$$

If we further require the "discovery" criterion for the background fluctuation which is represented by v^{n_s-1} to be small, we arrive at the result of a "Boon of Dimensionality":

$$v^{n_s-1} < p_{max} \tag{3}$$

Note that for a multi-dimensional Gaussian signal

$$v \approx \prod_{i=1}^{D} \sigma_i, \tag{4}$$

where σ_i is the Gaussian width in each dimension.

Finally, note also that n_s should be larger than 1, and the number of signal events is always proportional to the size of the sample, i.e.,

$$n_s \propto N. \tag{5}$$

4 Application to Particle Physics

Let us consider the search for supersymmetric particles, discovery of which would establish "new physics", i.e., physics beyond the Standard Model. Supersymmetric particles are thought to be produced in pairs, if "R-parity" is conserved, and their masses are expected to be in the TeV range. If a pair of such particles is created, each will decay [9] further to a single SUSY particle and a SM particle. As described above, such a decay chain continues until the LSP escapes the detector, or not if its decay violates R-parity. It is conceivable then that a chain of three or more SUSY particles could be detected via decays to SM particles.

The observables would be the three SUSY particle masses m_1, m_2, and m_3. If we think of a search in the three dimensions of such data, the signal would be a cluster in the 3-dimensional space of masses while a slowly varying background can be expected, which typically decreases with mass. Such a distribution, necessarily background dominated due to lack of observation thus far, can be transformed into a more-or-less uniform distribution and scaled to lie in the unit hypercube (here D = 3, so it is simply a cube).

Then all the statistical arguments given above apply and the signal would occupy a volume determined by the width of the signal mass peaks in each dimension per Eq. 4. If v is sufficiently small, i.e., the Gaussian widths after transformation into the hypercube are sufficiently narrow, Eq. 3 applies and discovery may follow. Of course, discovery depends on the size of the signal, n_s.

5 Further Thoughts

The discovery of small signals described above depends on the signal being clustered and the background widely distributed in the search space. There are other ways of finding clusters, and the literature on cluster-finding is quite large [10]. In particular, one may consider a recent technique which describes a recursive product of spacings (RPS) statistic which is touted to be better than usual techniques [11] or other ways to find signals in data.

6 Conclusion

In summary, we have established a criterion for discovery of a clustered signal in many dimensions by turning the well-known "curse of dimensionality" in high dimensions into a "boon of dimensionality" that helps to defeat large backgrounds in big data analysis. An application of this in discovery problems, such as the problem of discovery of supersymmetry (SUSY) in fundamental physics is described.

Acknowledgments. Many thanks to the organizers Profs. Bhalla, Sachdeva, and Watanobe, and the crew of BASE22 at NIT Delhi, India and at the University of Aizu, Japan.

References

1. Barnes, V.E., et al.: Observation of a hyperon with strangeness minus three. Phys. Rev. Lett. **12**(8), 204 (1964)
2. Aitchison, I., Hey, A.: The Standard Model describes all known forces of physics excluding gravity and results from the work of many physicists. A good introductory text is "Gauge Theories in Particle Physics: A Practical Introduction". Institute of Physics (2003). ISBN: 978-0-585-44550-2
3. Wess, J., Bagger, J.: Supersymmetry and Supergravity. Princeton Series in Physics, 2nd edn. Princeton University Press, Princeton (1991)
4. Cabrera, B.: First results from a superconductive detector for moving magnetic monopoles. Phys. Rev. Lett. **48**(20), 1378–1381 (1982)
5. Krizan, P.: "Belle II at SuperKEKB Status and Plans", talk given at "LHC Days at Split", October 2022. arXiV:2207.11275 [hep-ex]
6. ATLAS Collaboration: Observation of a new particle in the search for the Standard Model Higgs boson with the ATLAS detector at the LHC. Phys. Lett. B **716**, 1–29 (2012)
7. Bellman, R.E., Rand Corporation: Dynamic Programming, p. ix. Princeton University Press (1957). ISBN: 978-0-691-07951-6
8. Bellman, R.E.: Dynamic Programming. Courier Dover Publications (2003). ISBN: 978-0-486-42809-3
9. Canepa, A.: Searches for supersymmetry at the Large Hadron Collider. Rev. Phys. **4**, 100033 (2019)
10. Bouveyron, C., Celeux, G., Murphy, T.B., Rafter, A.E.: Model-Based Clustering and Classification for Data Science: With Applications in R. Cambridge University Press, Cambridge (2019)
11. Eller, P., Shtembari, L.: A goodness-of-fit test based on a recursive product of spacings. arXiv:2111.02252 [stat.ME]

Neuro-Symbolic Regression
with Applications

Nour Makke[✉] and Sanjay Chawla

Qatar Computing Research Institute, HBKU, Doha, Qatar
{nmakke,schawla}@hbku.edu.qa

Abstract. Discovering symbolic models is growing in popularity with
the increasing interest in interpretable machine learning. Symbolic
regression is the task of learning an analytical form of underlying models
in data. Two machine learning techniques have proven their effectiveness:
reinforce trick and transformer neural network. This paper discusses in
detail the two techniques and presents the application of symbolic regression on a simulated data set that describes a high-energy physics process.

Keywords: Model discovery · Symbolic regression · Neural network ·
Transformer network · Physics data

1 Introduction

Model discovery in a data-driven manner is a standard task of machine learning
(ML). Models learned from data often capture hidden patterns and can be used
to make accurate predictions associated with the studied phenomenon. Based
upon the technique adopted in the learning process, models are categorized as
follows: uninterpretable or "BlackBox" models such as deep neural networks,
where the relationship between the input and the output is neither transparent
nor tractable, and interpretable or "Whitebox" models such as decision tree,
where the input-output relationship is accessible, and allows for reasoning.

While there has been a lot of success in ML-based models in making highly
accurate predictions, they remain uninterpretable and opaque. The increasing
need for interpretable ML, especially in critical disciplines, motivates the development of ML-based methods that are predictive and interpretable. This is especially important for the application of ML in physical sciences, which is needed
more than ever with the tremendous amount of data collected. For example, in
high-energy physics experiments, a large amount of data is generated, and it is
not humanly possible to carry out a manual examination to look for patterns. In
such a scenario, a machine learning model effectively summarizes the data and
can be used for making predictions. The model can then be introspected to elicit
the prediction process if the predictions are accurate. More generally, physics is
essentially described by mathematical equations, and one could ask if we can
learn such equations directly from data. This is the symbolic regression. It was
introduced back in 1970 [1]. It came back in 2009 with the commercial platform

© The Author(s), under exclusive license to Springer Nature Switzerland AG 2023
S. Sachdeva et al. (Eds.): BDA 2022, LNCS 13830, pp. 38–50, 2023.
https://doi.org/10.1007/978-3-031-28350-5_4

Eureqa [2], which was publicized as a scientific discovery tool and developed using an evolutionary algorithm called genetic programming (GP). SR has since then been developed within the GP community. More recently, symbolic regression has been further tackled with deep learning-based tools. The last decade has witnessed the revolution of the deep learning field triggered by the development of ImageNet [3], with a recent review in [4]. The enormous growth of deep learning was based on the potential of deep neural networks, being universal function approximators particularly known for being end-to-end differentiable in their free parameters, predictive, and highly accurate.

The extensive use of deep neural networks over the last decade has unveiled the limitations of pure data-centric machine learning methods, ranging from adversarial attacks to explainability and fairness. These limits have triggered a trend to incorporate abstract concepts into the machine learning framework, such as graphs and symbolic representations, which help capture the complexity hidden in data while preserving interpretability. Recent machine learning methods aiming at avoiding out-of-distributions effects and promoting interpretability are growing in popularity. In particular, symbolic regression is attracting a wider research community within the machine learning field. In contrast, its adoption in scientific disciplines is at a very early stage, mainly due to immature developments made in this sub-field. Symbolic regression problems can be addressed using various approaches; see [5] for a full review. This paper discusses in detail and compares both the reinforce trick, which is used in reinforcement learning problems and the transformer network approach, which solves symbolic regression problems analogously to a language translation task. These approaches are of particular interest and have proven their effectiveness.

In the following, we present the definition of a function class (Sect. 2.1) and an expression tree representation (Sect. 2.2) within the problem statement section (Sect. 2). Furthermore, we discuss the symbolic regression problem within the zeroth-order optimization context (Sect. 3) and the attention mechanism context (Sect. 4). Finally, we present two state-of-the-art symbolic regression methods and compare their results on a physics application in Sect. 5. The conclusion is given in Sect. 6.

2 Problem Statement

Given a data set $\mathcal{D} = \{x_i, y_i\}$, where $x_i \in \mathbb{R}^d$ and $y_i \in \mathbb{R}$, symbolic regression aims at finding the mapping $(f^\star)$ in the class of mappings $\mathcal{F} : (f : \mathbb{R}^d \to \mathbb{R})$ that minimizes the loss function as follows:

$$f^\star = \arg\min_{f \in \mathcal{F}} l(f) \tag{1}$$

where f represents a non-linear, expressive, and parameterized function (e.g., neural network), and the loss function (l) is defined by:

$$l(f) = \sum_{i=1}^{b} l(f(x_i), y_i) \tag{2}$$

2.1   Function Class

The class of function $\mathcal{F}$ can be defined as the set of functions that can be obtained by the composition of mathematical operations and operands from a pre-defined library $\mathcal{L}$. The latter includes unary and binary mathematical operations, arithmetic operations, variables, and placeholders. Also, it can be extended to include as many operations as needed, as follows:

$$\mathcal{L} = \{+(\cdot,\cdot), -(\cdot,\cdot), \times(\cdot,\cdot), \div(\cdot,\cdot), \cos(\cdot), \sin(\cdot), \tan(\cdot), \exp(\cdot), \log(\cdot), \sqrt{(\cdot)}, x, etc.\}$$

The class of function $\mathcal{F}$ defines the search space in the symbolic regression problem and is, by definition, of discrete nature.

2.2   Expression Tree Representation

Every mathematical equation can be represented as a unary-binary tree. The latter is a rooted tree in which internal nodes are mathematical operators (e.g., $\div$, log) and terminal nodes are operands, i.e., input variable or constant (e.g., x). Operators may be unary (one argument, e.g., cosine) or binary (two arguments, e.g., addition). For example, the function $f(x) = x^2 - \cos(x/y)$ is represented by the tree illustrated in Fig. 1.

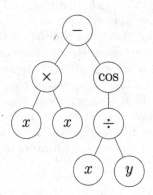

Fig. 1. Unary-binary tree-like structure of the mathematical function $f(x,y) = x^2 - \cos(x/y)$.

Furthermore, an expression tree can be represented as a unique sequence of symbolic representations, following the polish notation [7], by traversing the (binary) tree top to bottom and left to right in a depth-first manner.

$-$	$\times$	x	x	cos	$\div$	x	y

3   Zeroth-Order Optimization Problem

Symbolic regression methods aim to learn a mapping from a set of input-output pairs of numerical values to an analytical equation by minimizing a loss function. Therefore, SR learns the structure and the parameters of underlying models in data. Similarly to ML-based methods, the optimization task involves, through gradient-descent approaches, the computation of gradients of the loss function. However, the latter is not differentiable, which makes it a zeroth-order optimization problem [8].

A state-of-the-art symbolic regression application is Deep Symbolic Regression (DSR) [9], in which mathematical equations are represented by symbolic expression trees. DSR uses a deep neural network to generate sequences and the "Reinforce trick" [6] to train it. The sequence generator is chosen to be a recurrent neural network (RNN). The latter is a parameterized distribution over mathematical expressions $p(\tau|\theta)$ that allows backpropagation of a differentiable loss function with respect to parameters θ. Symbols of a sequence τ are generated one at a time, and each symbol τ_i is sampled from a pre-defined library of mathematical operations, e.g., $\mathcal{L} = \{+, -, \times, \div, \sin, \cos, \log, x\}$. For each symbol τ_i, RNN takes as input the parent and sibling nodes of the symbol being sampled and outputs a probability distribution over $\mathcal{L}$, conditioned by the preceding symbols $\tau_1, \cdots, \tau_{(i-1)}$, as illustrated in Fig. 2. RNN is trained using the reinforcement learning-based technique described in the following. Once the mathematical expression is sampled, it is evaluated with a reward function $R(\tau)$ that is defined using the normalized root-mean-square error (RMS), $R(\tau) = 1/(1 + \text{RMS})$.

In this approach, the optimization problem reduces to maximize the reward function. For this goal, the standard policy gradient objective defined by the expectation of the reward is considered, i.e., $J(\theta) = \mathbb{E}_{\tau \sim p(\tau|\theta)}[R(\tau)]$. The optimization problem is thus formulated as:

$$\theta^* = \arg\max_\theta J(\theta) \tag{3}$$

This optimization task (Eq. 3) is challenging because the reward function $R(\tau)$ is not differentiable with respect to learnable parameters θ. Here the "Reinforce trick", i.e., REINFORCE, originally introduced in the reinforcement learning community [6] is used. It transforms the gradient of the reward $\nabla_\theta R(\tau)$ to the gradient of the logarithm of the policy $\log(p(\tau, \theta))$ as follows:

$$\nabla_\theta \mathbb{E}_{p(\tau,\theta)}[R(\tau)] = \nabla_\theta \int R(\tau)p(\tau, \theta)d\theta$$

$$= \int R(\tau)\nabla_\theta p(\tau, \theta)d\theta$$

$$= \int R(\tau)\frac{\nabla_\theta p(\tau, \theta)}{p(\tau, \theta)}p(\tau, \theta)d\theta \tag{4}$$

$$= \int R(\tau)\nabla_\theta \log(p(\tau, \theta))p(\tau, \theta)d\theta$$

$$= \mathbb{E}_{\tau \sim p(\tau,\theta)}[R(\tau)\nabla_\theta \log p(\tau, \theta)]$$

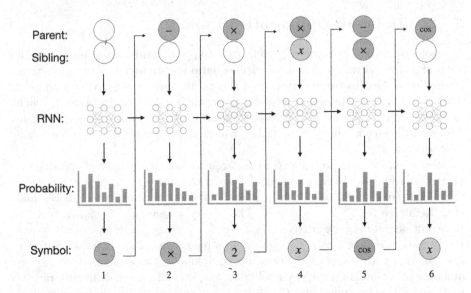

Fig. 2. An example of generating the mathematical expression $f(x) = 2x - \cos(x)$ from RNN. For the first node of the (binary) tree, empty symbols are given to the RNN as input because the tree node does not have a parent or sibling. Following the pre-order traversal of the tree, symbols are autoregressively sampled until the tree is completed.

The importance of this result is that $\nabla_\theta J(\theta)$ can be evaluated by computing the sample mean over a batch of N sampled expressions as follows:

$$\nabla_\theta J(\theta) \approx \frac{1}{N} \sum_{i=1}^{N} R(\tau^{(i)}) \nabla_\theta \log p(\tau^{(i)}|\theta) \tag{5}$$

This, in turn, can be optimized using gradient ascent:

$$\theta \leftarrow \theta + \alpha R(\tau) \sum_{i} \nabla_\theta [\log p(\tau, \theta)] \tag{6}$$

A key technique that boosts the performance of DSR is the use of "risk-seeking policy gradient", i.e., to optimize the best-case performance of a policy instead of optimizing its average performance. For this goal, the top-performing ϵ fraction of expressions found during training are selected, and a new learning objective is defined by:

$$J(\theta, \epsilon) = \mathbb{E}_{\tau \sim p(\tau|\theta)} [R(\tau) \mid R(\tau) \geq R_\epsilon(\theta)] \tag{7}$$

The Reinforce trick is used to estimate the new objective function where only the top ϵ fraction of expressions from each batch are used in the gradient computation.

4   Transformer Neural Network for Symbolic Regression

A transformer neural network (TNN) is a novel neural network architecture designed to treat sequential data (e.g., translation tasks). It is a sequence-to-sequence model that was developed in Natural Language Processing (NLP) [11] community to capture and model long-range dependencies in sequential data based on the attention mechanism. Consider the English-to-French translation of the two following sentences (sequences of words):

En: The mouse did not eat the cheese because **it** was <u>sick</u>.
Fr: *La souris n'a pas mangé le formage parce qu'**elle** était malade.*

En: The mouse did not eat the cheese because **it** was expired.
Fr: *La souris n'a pas mangé le formage parce qu'**il** était périmé.*

The translation of the word "**it**" is different because of the different contexts in the two sentences. However, they are almost identical. The only difference between the two sentences is in the last word, which refers to the mouse in the first sentence (i.e., "**sick**"), whereas it refers to the cheese in the second sentence (i.e., "**expired**"). This is what the attention mechanism is about. It pays particular attention to the terms (of the sequence) with high weights. In this example, the noun that the adjective of each sentence refers to has a significant weight and is therefore considered for translating the word "it". Technically, an embedding x_i is assigned to each element of the input sequence, and a set of m key-value pairs is defined, i.e., $\mathcal{S} = \{(k_1, v_1), \cdots, (k_m, v_m)\}$. For each query, the attention mechanism computes a linear combination of values $\sum_j \omega_j v_j$, where the attention weights ($\omega_j \propto q \cdot k_j$) are derived using the dot product between the query (q) and all keys (k_j), as follows:

$$\text{Attention}(q, \mathcal{S}) = \sum_j \sigma(q \cdot k_j) v_j \qquad (8)$$

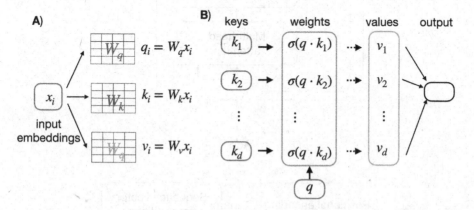

Fig. 3. A) The original embeddings (queries $\{q_i\}$, keys $\{k_i\}$, values $\{v_i\}$) computed from the embeddings of the input sequence $\{x_i\}$. **B)** Evaluation of Attention($q, \mathcal{S}$) (Eq. 8) for a query q.

Here, $q = xW_q$ is a query, $k_i = x_iW_k$ is a key, $v_i = x_iW_v$ is a value, and W_q, W_k, W_v are learnable parameters. The architecture of the self-attention mechanism is illustrated in Fig. 3.

For symbolic regression tasks, mathematical equations are regarded as sequences of symbolic representations, and TNN is used as a set-to-sequence model. Consider the function $f(x, y) = x^2 - \cos(x/y)$ whose tree is illustrated in Fig. 1. Its sequence of embeddings is given by:

$$x_1 : - \quad x_2 : \times \quad x_3 : x \quad x_4 : x \quad x_5 : \cos \quad x_6 : \div \quad x_7 : x \quad x_8 : y$$

In this example, for the prediction of the query $(x_8 : y)$, the attention mechanism will compute a higher weight for the binary division operator $(x_6 : \div)$ rather than for the subtraction operator $(x_1 : -)$ or the variables $(x_3 : x)$ or $(x_4 : x)$.

Transformers have an encoder-decoder structure. The structure of each block mainly comprises an attention layer and a feed-forward neural network. As an example, the encoder block of TNN is illustrated in Fig. 4. TNN takes the sequence of embeddings $\{x_i\}$ and outputs a "context-dependent" sequence of embeddings $\{y_i\}$, through a latent representation. The output can have a multipurpose use (e.g., translation task, classification task, etc.).

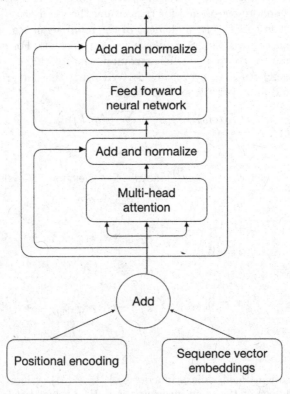

Fig. 4. Structure of the encoder in a transformer neural network. It comprises an attention layer and a feed-forward neural network [13].

"Neural Symbolic Regression that Scales" (NeSymReS) [12] is a recently developed TNN-based symbolic regression method. It builds on and combines two approaches that have proven to be highly efficient. The first is pre-training large models on large datasets, and the second is using transformers for tasks that involve symbolic operations. Whereas existing TNN-based methods use transformers in a pure symbolic domain (e.g., symbolic integration, solving differential equations, etc.), this approach aims at mapping numerical values to symbolic equations. Therefore it represents a set-to-sequence model that predicts an equation (expression tree) from a set of input-output pairs of numerical values (X, Y).

NeSymReS comprises two phases. First is the pre-training phase. A large amount of training data is generated using computers, which can generate an infinite amount of data with perfect accuracy at almost no cost. A training example is defined by $\{x_i, f(x_i)\}_{i=1}^n$ and an equation (τ) that represents the mapping $f : \mathbb{R}^{d_x} \to \mathbb{R}^{d_y}$. In the pre-training phase, the encoder maps each sequence $\{x, y\} \in (X, Y)$ into a latent representation z. The decoder then generates a sequence of symbols $\tau = \{\tau_1, \cdots, \tau_{|\tau|}\}$, where $|\tau|$ is the length of the sequence (traversal of the expression tree). The decoder outputs a probability distribution $P(\tau_i | \tau_{1:(i-1)}, z)$ given the latent representation z and the previously sampled symbols. Only mathematical operators (in the sequence) are sampled, i.e., numerical constants are replaced by a placeholder symbol $(\triangle)$, and will be fit at a later stage. This is called the skeleton. Each placeholder will be treated as an independent parameter in case of occurrence.

An example of sampling the mathematical expression $f(x) = 2x - \cos(x)$ in NeSymReS is illustrated in Fig. 5. This method first predicts the sequence of symbols as shown below, and then fits the numerical values of existing constants.

$$x_1 : - \quad x_2 : \times \quad x_3 : \triangle \quad x_4 : x \quad x_5 : \cos \quad x_6 : x$$

The second phase is the test time, in which the efficiency is measured on the test data set. A key factor in this approach is that it improves over time with data, and the (encoder-decoder) networks do not have to be retrained from scratch for each new experiment (equation), in contrast with all SR approaches, in particular, DSR.

5 Physics Application

This section presents the application of SR approaches to high-energy physics. A problem of interest is the so-called "hadronization mechanism" [14], which describes the formation of hadrons (such as protons and neutrons) from elementary particles (the so-called "quarks"). This process is studied using experimental measurements of particle distributions, which are observed to follow an exponential functional form. Experimental measurements are commonly performed in a multidimensional scheme for an in-depth understanding of the target phenomena. In our discussion, we will only consider two dimensions for simplicity, i.e., $\mathcal{D}(x, y)$. $\mathcal{D}$ represents the distributions of particles as a function of one variable

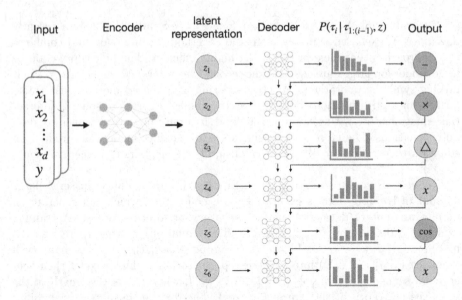

Fig. 5. An example of sampling the mathematical expression $f(x) = 2x - \cos(x)$ in "Neural Symbolic Regression that Scales" application based on transformers. The triangle symbol represents a numerical constant that will be fit at a later stage. $P(\tau_i|\tau_{1:(i-1)}, z)$ represents the probability distribution over the symbol τ_i given previous symbols $\tau_{1:(i-1)}$ and the latent representation z.

x, in ranges of another variable y. The goal of such measurements is to learn the analytical form of the underlying mechanism in terms of x and y. This goal is traditionally achieved by fitting data points using a pre-defined functional form with unknown parameters, which is not an ideal solution simply because the underlying model is unknown. This application aims at applying symbolic regression to such data points, and we expect that the symbolic models extracted in the multiple ranges of y will be the same, however, numerical values of the parameters are expected to change from one range of y to another.

In this application, we will use simulated data points instead of experimental ones. Assume that experimental measurements follow the functional form:

$$f(x, y) = \frac{1}{a(y)} \exp\left(\frac{-x}{a(y)}\right) \qquad (9)$$

Here x and y represent two physical properties of the particles (e.g., energy and momentum), and a is a fit parameter. The function $f(x)$ exhibits an implicit dependence upon y via its free parameter a. To imitate experimental data, we simulate five data sets, each corresponding to a different range of y, i.e., to a different value of a in the simulation. Figure 6 illustrates the function f as a function of x ($\in [0, 3]$) in intervals of y. The slopes of the curves significantly change for different intervals of y, up to six orders of magnitude difference at

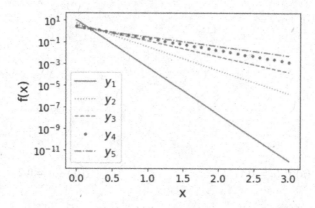

Fig. 6. Mathematical expression $f(x,y) = 1/a(y)\exp(-x/a(y))$ as a function of x in different ranges of y. Each range of y corresponds to a different value of a: y_1 ($a = 0.1$, solid curve), y_2 ($a = 0.2$, dotted curve), y_3 ($a = 0.3$, dashed curve), y_4 ($a = 0.4$, loosely dotted curve) and y_5 ($a = 0.5$, dash-dotted curve).

high values of x between the smallest and the largest intervals of y, i.e., between the smallest and the largest values of a given that data points are simulated.

Simulated data sets, i.e., $\mathcal{D} = \{x_i, f(x_i)\}_{i=1}^n$, each consisting of 30 input-output pairs, are generated with perfect accuracy, and no noise is added. In this application, we will apply the SR approaches DSR and NeSymReS, and compare their results. Ideally, we expect to learn the same underlying model, i.e., $f(x) = 1/a\exp(-x/a)$, in the five intervals of y, while obtaining different values of a. We use a pre-trained version of NeSymReS, whereas DSR is trained for each new experiment (data set), as previously mentioned in the text. A total of 20 runs is performed on each experiment using NeSymReS and 100 runs using DSR.

Figure 7 shows simulated data points (solid markers) and the results obtained by applying both SR approaches to simulated data sets (curves). The resulting equations and recovery rates (how many times the exact mathematical equation is covered) are presented in Table 1 for each experiment. Results obtained using NeSymReS show a significant discrepancy between the predicted expressions and the ground-truth function for all intervals of y except one. An in-depth investigation shows that, in a successful experiment, either the skeleton is correctly predicted ($c_1 * \exp(c_2 * x)$ in this case), or it is not accurately predicted, but, it reduces to the correct equation after fitting the constants. For example, for the only successful case in this application, i.e., $y_1(a = 0.1)$, NeSymReS predicts the following skeleton:

$$c_1 * (c_2 * x + \exp(c_3 * x))^{\text{pow}}$$

where c_1, c_2, and c_3 are fit constants, and pow is a power coefficient. They have the following values:

$$c_1 \approx 10 \quad c_2 = 2.075 \times 10^{-6} \quad c_3 \approx 10 \quad \text{pow} = -1$$

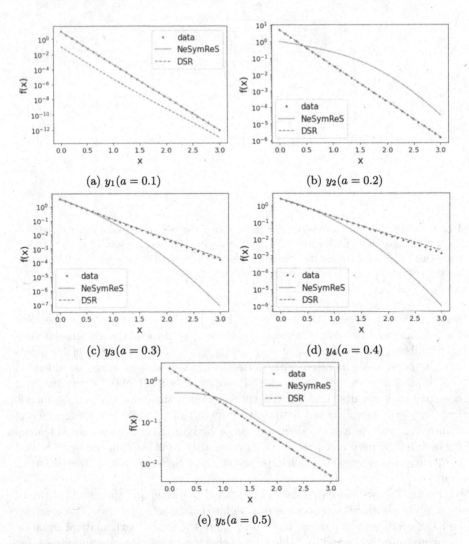

(a) $y_1(a = 0.1)$

(b) $y_2(a = 0.2)$

(c) $y_3(a = 0.3)$

(d) $y_4(a = 0.4)$

(e) $y_5(a = 0.5)$

Fig. 7. Results of DSR and NeSymReS applied to simulated data. The functional form $f(x, y) = 1/a(y) \exp(-x/a(y))$ is used to generate five data sets with $x \in [0, 3]$ in five ranges of the variable y.

The predicted expression becomes:

$$10 * (0 * x + \exp(10 * x))^{-1} = 10 * \exp(-10 * x)$$

Therefore the predicted equation is equal to the ground-truth function. Consider one of the failed experiments, i.e., $y_2(a = 0.2)$. NeSymReS predicts the following expression:

$$(x + \exp(x)^2)^{-1}$$

In this case, the NeSymReS predicts a skeleton that does not match the ground-truth function and does not include any constant to fit, in contrast with the successful case. In conclusion, the main reason for failure, in this method, is the incorrect skeleton prediction rather than at the level of fitting the numerical values of the constants in the predicted skeleton.

DSR successfully predicts mathematical equations for two (out of five) data sets ($a = 0.2$ and $a = 0.5$), and it produces an accurate fit for the other ranges of y with an R-coefficient that is greater than 0.99. Consider one successful experiment, for example, $y = 0.5$. DSR predicts, in this case, a sequence of 14 symbols given by:

÷	+	cos	−	x	x	cos	−	x	x	exp	+	x	x

which is equivalent to:

$$\frac{\cos(x-x) + \cos(x-x)}{\exp(x+x)} = \frac{2}{\exp(2*x)} = 2*\exp(2*x)^{-1}$$

Therefore, equations predicted by DST do not necessarily match the exact sequence of the ground-truth function.

In summary, the success rates for both methods are less than 50% in this application. However, the predictions made by DSR fit very well the data for the whole range of x in all ranges of y, even in cases when the function is not correctly predicted, in contrast to NeSymReS, although the latter has been shown to outperform DSR on various data benchmarks in terms of test-time compute and out-of-distribution prediction [12].

Table 1. Equations and recovery rates obtained by the application of SR methods, DSR and NeSymReS, on simulated data sets. The equation predicted in y_1-range is $f(x) = e^{-10*x+e^{-\cos(x)}-e^{e^{-\sin(e^{-4e^{-6e^{-1e^{x}}}})}}}$, and the equation predicted in y_2-range is $f(x,y_3) = e^{-3*x+e^{-e^{-x}}+e^{-e^{x1-e^{\cos(e^{-x})}}}}$.

Method →	DSR			NeSymReS		
y-range ↓	Fit	$f(x)$	Rate (%)	Fit	$f(x)$	Rate (%)
$y_1 (a=0.1)$	✗	In caption	0	✓	$10e^{-10x}$	100
$y_2 (a=0.2)$	✓	$5e^{-5x}$	5	✗	$(x+e^{2x})^{-1.15}$	0
$y_3 (a=0.3)$	✗	In caption	0	✗	$3.3\,e^{(-x^2-2.8x)}$	0
$y_4 (a=0.4)$	✗	$e^{-2x-\cos(e^{e^{-3/8x}})}$	0	✗	$2.4\,e^{(-x^2-1.9x)}$	0
$y_5 (a=0.5)$	✓	$2e^{-2x}$	100	✗	$\frac{x}{(x^5+2x)}$	0

6 Conclusion

Learning interpretable models is growing in interest and in popularity in the last decade. Symbolic regression represents a key method to learn interpretable models in a purely data-driven manner. Recent developments in the symbolic regression field have shown that the use of deep neural networks boosts the performance of these methods. In this paper, we presented the application of two symbolic regression methods on simulated data towards applying them to experimentally measured data, and we showed that state-of-the-art symbolic regression methods do not necessarily produce consistent results, as one expects.

References

1. Langley, P.: Data-driven discovery of physical laws. Cogn. Sci. **5**(1), 31–54 (1981). https://doi.org/10.1016/S0364-0213(81)80025-0. Conference 2016
2. Dubcakova, R.: Eureqa: software review. Genet. Program Evolvable Mach. **12**(2), 173–178 (2011). https://doi.org/10.1007/s10710-010-9124-z
3. Deng, J., Dong, W., Socher, R., Li, L.-J., Li, K., Fei-Fei, L.: ImageNet: a large-scale hierarchical image database. In: 2009 IEEE Conference on Computer Vision and Pattern Recognition, pp. 248–255 (2009). https://doi.org/10.1109/CVPR.2009.5206848
4. Chawla, S., et al.: Ten years after ImageNet: a 360° perspective on AI. arXiv: 2210.01797
5. Makke, N., Chawla, S.: Interpretable scientific discovery with symbolic regression: a review. arXiv:2211.10873
6. Ronald, W.J.: Simple statistical gradient-following algorithms for connectionist reinforcement learning. Mach. Learn. **8**, 229–256 (1992). https://doi.org/10.1007/BF00992696
7. Lukasiewicz, J.: Aristotle's Syllogistic from the Standpoint of Modern Formal Logic, Second Edition Enlarged, pp. xvi 222. Clarendon Press, Oxford (1957). Cloth, 305.net
8. Liu, S., Chen, P., Kailkhura, B., Zhang, G., Hero III, A.O., Varshney, P.: A primer on zeroth-order optimization in signal processing and machine (2020). arXiv:2006.06224
9. Petersen, B.: Deep symbolic regression: recovering mathematical expressions from data via policy gradients. CoRR (2019). arXiv:1912.04871
10. Mundhenk, T.N., Landajuela, M., Glatt, R., Santiago, C.P., Faissol, D.M., Petersen, P.K.: Symbolic regression via neural-guided genetic programming population seeding. CoRR (2021). arXiv:2111.00053
11. Vaswani, A., et al.: Attention is all you need. CoRR (2017). arXiv:1706.03762
12. Biggio, L., Bendinelli, T., Neitz, A., Lucchi, A., Parascandolo, G.: Neural symbolic regression that scales (2021). arXiv:2106.06427
13. Wood, T.: Transformer neural network. https://deepai.org/machine-learning-glossary-and-terms/transformer-neural-network
14. Andersson, B., Gustafson, G., Soderberg, B.: A general model for jet fragmentation. Z. Phys. C **20**, 317 (1983). https://doi.org/10.1007/BF01407824

Data Science: Architectures

Introducing Federated Learning into Internet of Things Ecosystems – Maintaining Cooperation Between Competing Parties

Karolina Bogacka[1], Anastasiya Danilenka[2],
Katarzyna Wasielewska-Michniewska[1(✉)], Marcin Paprzycki[1],
Maria Ganzha[2], Eduardo Garro[3], and Lambis Tassakos[4]

[1] Systems Research Institute, Polish Academy of Sciences, Warsaw, Poland
{k.bogacka,k.wasielewska,m.paprzycki}@ibspan.waw.pl
[2] Warsaw University of Technology, Warsaw, Poland
{anastasiya.danilenka,maria.ganzha}@pw.edu.pl
[3] TwoTronic Gmbh, Meitingen, Germany
[4] Prodevelop, Valencia, Spain

Abstract. In practical realizations of a Federated Learning ecosystems, the parties cooperating during the training process, and that later use the trained/global model may consist of competing institutions. This can result in incentives for malicious behavior, which can infringe on the safety and data privacy of other participants. Additionally, even in cases devoid of foul play, the format of the data stored locally, and the equipment available for training, may differ between participating institutions. This necessitates creation of a flexible and adaptable preprocessing pipeline, including a comprehensive registration and data preparation process. Among others, it should identify the affiliation of the joining device(s), maintain appropriate data privacy mechanisms, and compensate for the heterogeneity of the devices that are to participate in model training. In this context, the practical aspects of deploying federated learning solutions, in real-life production environments, are discussed.

Keywords: Federated learning · Internet of Things · Data privacy · Business requirements · Coopetition

1 Introduction

One of the well-known problems in applying Machine Learning (ML) lies in the cost and time required to collect, consistently label, and "normalize" large

Work supported by ASSIST-IoT project funded from the European Union's H2020 RIA program under grant 957258. Work of Maria Ganzha, Anastasiya Danilenka and Karolina Bogacka was funded in part by the Centre for Priority Research Area Artificial Intelligence and Robotics of Warsaw University of Technology within the Excellence Initiative: Research University (IDUB) programme.

datasets. Moreover, even if the overall volume of the existing data is large, labeled, and in the same format, such dataset can still be "split" between different stakeholders, who do not want to and/or cannot share their data [32]. A typical example of such situation occurs when medical data belong to different hospitals, or when market competitors want to benefit from a high-quality model but do not want to disclose their data (or its characteristics). Moreover, regardless of the push towards cloud-based solutions, hosting data in a centralized location still poses security risks [23,28,40].

To address these (and a few other [16,31]) challenges Federated Learning (FL) has been proposed [25]. In particular, it has been envisioned as a mechanism to support coopetition. Here, coopetition is understood as a business scenario in which multiple entities can compete in the business context, while cooperating when training a global/shared model.

Federated Learning, in its basic format, is envisioned as a process of *federating* results of local training, performed by (often heterogeneous) edge devices, known as clients. Usually, such a process is "synchronized" and "orchestrated" by a "central server". However, different topologies of FL ecosystems have been explored (see, for instance, [3,11,12,17,21,27] and Sect. 4). In a typical FL scenario, (i) clients train an ML model using local data; next, (ii) they send updated parameters to the server, which (iii) aggregates them, and (iv) updates the (central) shared model [29], later (v) shared model is sent back to the clients to continue training. This process is repeated until a global stopping criterion is met. Obviously, in this scenario, private data never leaves the clients [20]. A lot of research is devoted to the FL process itself, however, it is mostly implemented and tested using cloud-based platforms. Hence, important practical issues and real experiences are omitted [19,36]. Moreover, majority of FL-oriented research assumes that the training is "ready to be started", i.e. data is appropriately preprocessed, details of model to be trained have been agreed on (between business stakeholders) and the software needed to realize training has been deployed, etc. This observation provides the context to what follows.

The Internet of Things (IoT) is likely to become one of the key areas of FL application, following the trend to replace cloud-centric solutions with the edge-cloud continuum approaches [9]. This, in turn, is happening because sending data to, storing data in, and providing resources in a data center (i.e. central cloud, or cloud federation), may not be sustainable for large-scale complex IoT deployments, where data volume, transmission costs, and latency are of extreme importance. Hence, "computing" has to take place near/at the edge, i.e., physically close to sensors and/or users [15] and the resulting edge-cloud continuum ecosystem is seen as the necessary direction for the evolution of the Next-Generation Internet of Things [19]. This being the case, to infuse such a system with intelligence, FL becomes a natural choice, as it is based on the assumption that localized ML is applied and data is not (cannot be) transferred to the central repository [29].

Let us now go back to the, mentioned above, conceptual gap related to the steps that have to be undertaken before the actual federated learning process is initiated. Hence, let us assume that FL is to be instantiated "from the scratch" in a, newly deployed, IoT ecosystem. Here, the core of the FL job has to be agreed to, by the industrial IoT ecosystems stakeholders, i.e. what neural network is to be trained, what images are to be used in model training, what are system requirements for the clients, and what is the stopping criterion. Furthermore, a large number of practical infrastructure-related issues, recognized in [2], have to be addressed. They may include (a subset of) the following: (i) heterogeneity of clients and networks that can cause delays, or introduce "stragglers" (weaker/more busy clients), slowing down the training process or making use of "training rounds" difficult (if not impossible); (ii) resources on the (far) edge devices and their battery life, that tend to be limited, impeding the use of large models; (iii) data used for the training that can be highly redundant; (iv) clients that can suddenly drop out. Besides these issues, coming from the nature of the environment, there are other use cases and/or business-specific aspects that need to be considered. These include: (i) necessary data preprocessing, which is a crucial step for having an effective training pipeline and which may not be exactly the same for each node (e.g. image scaling and/or normalization), (ii) preservation of data and metadata privacy, (iii) security of communication, (iv) avoidance and detection of foul play by clients participating in the training process, and (v) costs of communication.

In this context, this contribution: (a) reflects the nature of practical challenges that actual FL deployments have to address "from the start", (b) shows how a reference architecture, proposed for Next-Generation IoT, supports FL deployment, and (c) illustrates how, in the proposed approach, mentioned business-specific challenges can be handled.

Hence, the remaining parts of this work are organized as follows. In Sect. 2 a practical IoT-based scenario from the ASSIST-IoT[1] project is described, introducing key requirements for "practical federated learning being realized in IoT ecosystems". In Sect. 3, pertinent state-of-the-art is summarized. Following, in Sect. 4, the reference FL architecture proposed in ASSIST-IoT project is briefly outlined. Next, in Sect. 5, the ways of adapting and using the proposed architecture so that the data preprocessing and privacy requirements are fulfilled are discussed. Finally, in Sect. 6, a summary of contributions, and directions for future work are provided.

2   Federated Learning Use Case in IoT Deployment

The use case presented in this work is provided within the scope of the ASSIST-IoT project, and considers a car damage detection and documentation task. The main goal is to provide fast and accurate damage recognition, based on the image(s) of a car. Source images come from the set of distinct scanners, which are fixed in their geolocation, and consists of a number of cameras each (Fig. 1a). Each scanner camera observes the car passing through the scanner gate from a different perspective/angle (front, left and right side, top, ground) and takes photos. Note that, on the basis of business requirements analysis, it is assumed

---

[1] https://assist-iot.eu/.

that scanner gates can belong to different stakeholders, be of varying technical specifications, and operate in varying conditions. The common goal of the stake-holders is to obtain a model that is as accurate as possible, but on the other hand they do not want to disclose their private data to their potential market competitors. Since, for instance, different scanners may belong to different car rental companies, this is a clear case of coopetition.

In terms of FL, either scanners, or individual camera,s can be treated as fed-erated clients (Fig. 1b). As data collected by the scanners may contain private information on the vehicle and the driver, and, moreover, participating compa-nies have their own obligations for data protection – data security, and overall privacy, are major concerns within this use case.

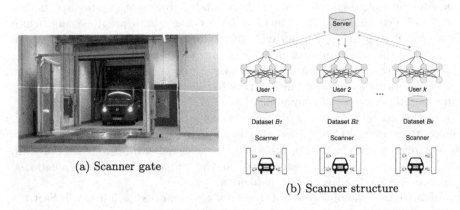

(a) Scanner gate

(b) Scanner structure

Fig. 1. Car damage recognition – scanner gate.

In the discussed use case, the main steps for the car damage detection task include: (i) car detection, within the image, (ii) finding a region of interest (ROI), e.g. bounding box with possible damage, (iii) performing segmentation to pre-cisely define the damaged region inside the ROI, (iv) classifying found damages, (v) reporting the results to the human operator and storing them in the database.

The nature of the task implies some data-dependent challenges that were also faced by researchers solving similar problems. The presence of the following data features may complicate the car damage detection task: (i) illumination and weather conditions (e.g. reflection can be mistaken for a scratch), (ii) background that features additional cars [22] that may not be the target during training step, (iii) image orientation and angles may confuse the model during the damage classification task, as the same damage, seen from a different perspective, can look different for the model [22] (iv) small objects (e.g. small scratches) are generally harder to annotate, detect and classify.

Another set of issues comes from the federated nature of the task. As the scan-ners can belong to different institutions, federated clients may become subjects to non-independently identically distributed (non-IID) local data. Thus, utilizing non-IID classification from [41], it can be stated that data on local nodes can suffer from label skew (certain types of damage can be over/underrepresented on nodes, or certain types/models of cars may appear in higher frequency, e.g., if scanners are installed in a car service of a particular manufacturer), attribute skew (scanner gates themselves can have a varying number of cameras with changing positions and be located in different physical places, e.g. outdoors or indoors and, therefore, be subject to varying external conditions) and temporal skew (for example, due to the different data storage regulations, adopted by the participating institutions, or different weather conditions over time: sunny, cloudy, etc.).

A specific type of issue comes from the domain of the task. As the goal is to accurately detect damages, car design itself can mislead the model into classifying specific design elements, e.g. car emblem, as damage. For example, decorative thin stripes, placed on the body of the car, may be classified as scratches if their existence is not addressed during the training step, or manufacturer's own logos can be misinterpreted.

The aforementioned issues put a lot of emphasis on the importance of the data preprocessing step, which is a part of the FL training process preparation. Moreover, the fact that collected images will belong to different stakeholders brings the need for the preservation of privacy. In this context, let us summarize the relevant state-of-the-art, found in the literature.

3   Related Work

3.1   Data Privacy in Federated Learning

Privacy-preserving Machine Learning (PPML), as a concept, aims at protecting data privacy. Usually this is achieved by employing dedicated methods, based on secure multi-party computation (SMC), homomorphic encryption (HE), differential privacy (DP) or a combination of all of those [10,39]. Privacy-preserving Federated Learning (PPFL) can be then understood as a particular subdomain of PPML, which emphasizes a training process conducted by a federation of distributed devices, each utilizing its own dataset [39]. Even though FL enables collaborative training of a machine learning model, without the need to move any sensitive data out of the local system, some information can still be reverse-engineered from the frequently exchanged model parameters [8]. AT the same time, application of privacy preserving mechanisms results in an overhead. Therefore, the main challenge of the privacy preservation lies in finding the right balance between preserving the privacy of local data, and being able to effectively utilizing it.

Various ways of applying privacy-preserving mechanisms to the FL process lead to different categories of methods. One of them is encryption-based PPFL, which focuses on the application of cryptographic techniques, like the previously mentioned HE, SMC, and secret sharing [39].

Within the HE scheme, arithmetic operations can be directly performed on encrypted data without decryption. As a consequence, in PPFL solutions based on HE, the clients send their updates (or weights) in an encrypted form. These updates are then aggregated and returned without the need for decryption. Although this method allows for significant protection of the clients' privacy, it leads to significant computational and communication overhead [30].

When the SMC is applied, multiple parties can compute a common function of interest in a collaborative manner. During training, FL clients can use a secure weighted average protocol to encrypt the weights of the model and wait for the server to calculate the weighted average on the encrypted data. This set of techniques does not lead to accuracy loss [8]. However, similarly to HE, it also causes communication and computational costs [39].

Secret sharing-based PPFL methods allow a group of mutually distrustful clients to compute a joint aggregate value, without revealing any information to each other, about their private data [7]. The secret can only be reconstructed when large enough number of shares is being combined, which can be then used for secure model updating. Unfortunately, methods from this class are the most vulnerable to the malicious clients.

An entirely different category of PPFL are the perturbation-based methods. Many of them are based on DP, employed either on a global, or on the local, level [39]. The basis for DP lies in the injection of a degree of random Gaussian noise into the calculation in order to protect the original data. On the global level, it means adding noise during the aggregation step, which protects the system from malicious clients, but not from a malicious server [26]. Some techniques have been developed that combine this approach with adaptive clipping of gradients, sent by the local model when it exceeds a set value, creating a successful combination of global and local DP [37]. In local DP, noise is added to the gradient of each training batch, or the norm of each data sample is clipped, which offers a stronger privacy guarantee. Both methods are computationally efficient but they often result in lower accuracy of the final model and/or longer training time (more training epochs are required). They also lead to challenging model fitting [5]. Additionally, additive and multiplicative perturbation-based PPFL methods either add random noise to weight updates or transform the local dataset into another space, respectively. The first method tends to preserve the statistical properties of the dataset and is simple in implementation. However, it may degrade data utility [1]. Although the second method provides better obfuscation of the original data, it can also be vulnerable to a malicious server [18].

Aside from hybrid approaches, the last remaining category of PPFL is PPFL based on anonymization. An important motivation for the development of this category stems from the degradation of data utility and model performance, inherent in the perturbation-based methods. Nevertheless, the selection of an anonymization scheme that provides sufficient protection from modern attacks, like mGAN-AI, can be challenging [35].

In summary, there exist multiple ways of securing the FL training process. The specific decision, which single one, or combination of which to use (and how rigorously), should be left to the stakeholders. However, regardless of the final deployment decision, either approach has to be supported and easy to deploy within the system.

3.2 Security in Federated Learning

Using FL in business scenarios introduces a set of potential additional security vulnerabilities. Those vulnerabilities stem from a few various sources, starting with the choice of the communication protocol. Exchanging updates between the clients and the server(s) involves a significant amount of communication. Regardless of the mechanisms employed in order to protect the private information included in that communication, non-secure communication channels still present a sizable security risk [28]. Here, Secure Socket Layer (SSL) protocol is frequently used to provide reliable services over the transportation layer protocol. It supports end-to-end secure communication for IP networks through employment authentication and encryption [33]. SSL is widely utilized for secure data transfers on the Internet [14]. Therefore, SSL encryption should be supported in the system and applied to all messages exchanged between clients and the server, in order to introduce an additional layer of protection, going beyond securing model data participating in the training process.

3.3 Data Transformations in Federated Learning

There is little research concerning the effective instatiation and use of preprocessing pipelines (e.g., image oriented, AI-relevant transformations) in FL. In most existing works, the data is assumed to be either already preprocessed on every FL client, or able to enact the exact same set of data transformations regardless of the local capabilities of the client. However, some of the data transformation methods do require specific computing capabilities of the system, often including a strong graphical subsystem with GPU hardware. Especially in the case of image segmentation, often used in order to enhance the results of a more sophisticated machine learning model, neural networks are often used [13]. This implies the existence not only of enough memory and storage to use such neural network, but also of a particular set of dependencies. Additionally, there is a wide variation of data transformation methods in use, from image resizing and outlier removal [6] to the employment of refined clustering methods for dimensionality scaling [24]. In order to increase the availability of difficult datasets for some domains, extraction and preprocessing methods can be grouped into predefined, open-source pipelines. This approach also allows for future usability, reproducibility, and extensibility [38]. This also means that all required transformations have to be easy to deploy in the ecosystem that is being assembled.

4   Assist-IoT Federated Learning Architecture

Let us now outline the proposed architectural approach to Federated Learning in Next Generation IoT ecosystems. Before proceeding, note that in [2] the support for different topologies and resistance to sudden user dropout, network connection with interruptions, or uneven grouping of clients were discussed. The proposed FL architecture is developed according to the Reference Architecture (RA) introduced in the ASSIST-IoT project, and motivated by real-life scenarios originating from four industrial pilots (one of them being the comprehensive car monitoring use case, described above). This RA is based on the concept of encapsulation, which is instantiated in the form of enablers [42]. Enablers are the cornerstone of ASSIST-IoT architecture, and respond to the realisation of a modular architecture. In essence, an enabler is a collection of software (and possibly hardware) components – running on computation nodes – that work together to deliver a specific functionality of a system. ASSIST-IoT enablers are not atomic but presented as a set of interconnected components, bound together in a single package. In particular, an enabler component is an artifact that can be viewed as an internal part of an enabler that performs some specific action, necessary to deliver the functionality of an enabler as a whole.

As it can be seen in Fig. 2, the proposed Assist-IoT FL architecture is formed by four enablers: *FL Orchestrator*, *FL Repository*, *FL Training Collector*, and *FL Local Operations*. The *FL Orchestrator* is responsible for the configuration propagation to the other enablers, workflow management, and control over the FL life cycle. It also acts as the entrance gate for human interactions. Moreover, the *FL Orchestrator* may control the FL training process and constraints related to, e.g., the minimum number of clients, system requirements, and necessary data transformations. Next, the *FL Repository* is a supplementary enabler for storing models, algorithms, transformations and any data needed in the FL process. Finally, the *FL Training Collectors* and the *FL Local Operations* act as FL servers and clients, respectively. They are used, in the constructed system, as communicating components, remaining in constant contact, according to the gRPC protocol, by utilizing functionalities implemented as a part of the Flower library [4]. In other words, the *FL Training Collector* possesses the capabilities of an FL centralized server, while the *FL Local Operations* (located on edge clients) has the abilities of an FL client, with the focus placed on loading dataset for training and training a local model. The *FL Local Operations* enabler is responsible for loading and preprocessing the right subset of local data, and setting up the local model. It may also include mechanisms related to privacy, such as data encryption or differential privacy [28, 40]. Keeping this in mind, on the following section we will describe how this architecture can be used to conceptually represent and assemble and IoT ecosystem.

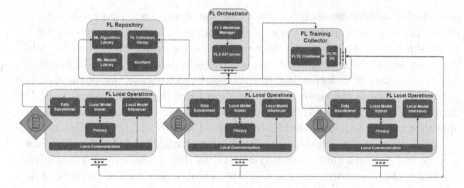

Fig. 2. Proposed FL in IoT architecture.

5 Federated Learning in IoT Ecosystems

5.1 Data Transformation

The data preparation step is a required stage in almost every ML task and, as has been discussed, it is also needed in FL pipelines. By design, in the proposed architecture, data preparation is realized by the *FL Local Operations* enablers, as the data is stored only on local devices. Moreover, as the local datasets, belonging to the corresponding clients, are expected to adhere to the same format at the start of the training, the data preprocessing definition is curated by the *FL Orchestrator*. Data transformation steps are defined during the task initialization phase and are distributed (from the *FL Orchestrator* to the *FL Local Operations*, together with the environment prerequisites. Therefore, every client is expected to meet the requested requirements, often by following the instructions in the given preprocessing routine. Here, note that different *FL Local Operations* may require different preprocessing. Therefore, the *FL Orchestrator* has to obtain all preprocessing modules (matching all local needs), before the FL ecosystem is realized. Here, note that some clients can suffer from limited memory size, computational ability, energy budget, or fail to meet other system requirements. In some situations these problems may not be known during system setup. Consequently, they may be evaluated as not fit for participating in training and dropped from the participants training list. This process is facilitated through interactions between *FL Orchestrator* and *FL Local Operations*.

In trivial cases, preprocessing routine consists of conventional operations that are used in standard ML tasks, and do not impose any difficulties upon the edge nodes, e.g. due to their ease of access and usage. For example, in order to use image processing operations like rotation, resizing, Gaussian noise removal, flipping, and others, only the installation of the library with given functions is needed. For instance, these operations can be used for the data augmentation step during the car damage detection [34], and further improve the outcomes of FL in limited local data scenarios. In the studied car damage detection use case, during requirements elicitation it has been determined that the basic image processing steps will include color scheme manipulation and image resizing.

Nevertheless, real-life tasks may request custom data preprocessing, which may increase the complexity of the preprocessing routine and add non-trivial steps to the data transformation. In the considered use case, one of the data preprocessing steps is the segmentation of a car from the image background. This data transformation operation cannot be performed by simple out-of-the-box operations, and needs an individually designed preprocessing step. For instance, the segmentation procedure can be implemented with the help of a supplementary neural network that is already trained to detect the required region from the image. Furthermore, using a separate model, solely for the purpose of data preprocessing, might be a challenging task in the presence of resource-restricted edge devices, as it requires additional memory, computational power, and a setup procedure for acquiring a supplementary model that may further face issues, for example, in data transmission. This means that (a) these restrictions have to be established during system design, and (b) they have to be checked during system deployment and operation.

Another requirement for the data preprocessing step, in the FL IoT pipeline, comes from the isolation of the clients. Here, by default, clients do not share information about statistical properties of their local datasets. It may restrict the usage of data preprocessing methods that rely on the characteristics of global data distribution, or require additional training, due to the possible presence of data non-IIDness. One more self-evident restriction is being placed on methods that require manual parameter tuning, as it is not possible to manually control the performance of such methods on the edge devices (it can be overcome by sending the results in terms of indexes without data content, which remains confidential). For the damage detection task, this limitation restricted the use of the circle Hough Transform, which allows the detection of circles on the image. This method could have helped to detect wheels and tires on the image. However, due to the variety of possible situations where the images were taken, in certain cases, parameter tuning was required, which is not desirable in terms of a fully automated data preprocessing pipeline.

All in all, preconditions for data transformation routine for IoT FL tasks may come from limited computational power, storage capacity, and other restrictions dictated by the edge devices themselves and the need for independent preprocessing operations. To deal with these problems, a dynamic mechanism for data transformation pipeline should be implemented. This solution should support use of custom transformations and be flexible as well as extensible. It should, therefore, ensure that support for a wide variety of ML libraries can be easily added to the system and subsequently used for data transformation. Additionally, major libraries like PyTorch or TensorFlow ought to be provided out-of-the-box for environments with relatively high computing capabilities. However, taking into account that the main goal of the approach is to deploy FL on the edge (in constrained environments), it can be assumed that only TensorFlow Lite should be available by default, due to its focus on achieving maximal efficiency of the training and testing process. Nevertheless, it can be assumed that the exact set of dependencies, supported by all *FL Local Operations*, is going to be decided before the setup of the system, by the stakeholders' consensus.

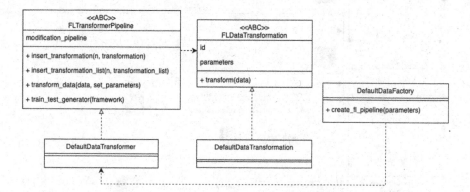

Fig. 3. Proposed software structure of the FL data transformation pipeline.

In order to allow for the effortless modification of the pipeline, the code structure is being constructed to be as abstract and universal as possible (see Fig. 3). Note that only a handful of methods will be necessary to define a new pipeline (class inheriting from *FLTransformerPipeline*) or data transformation (class inheriting from *FLDataTransformation*), all of them selected to be especially universal to supervised learning problems. A developer would be then able to create a new data transformation just by defining the appropriate unique *id* of the transformation, its input parameters, as well as placing the code necessary to conduct the transformation itself in the `transform` method (which can be of virtually any size or complexity). Such a simple format will allow the developer to quickly create a transformation object and then store it in the format of a *pickle file* in the *FL Repository*. Next, when needed, such object can be downloaded by all *FL Local Operations* instances that require it. A set of metadata describing the input parameters, and system requirements needed to run the transformation, should be stored along with the pickle file, together with a short description of the data transformation.

Furthermore, each data transformation pipeline will be built by forming a list of transformations in the order they should be applied by calling *insert_transformation* or *insert_transformation_list*. Here, a given dataset will be transformed in ways defined by the pipeline and the set of input parameter values. The dataset can be then shuffled, divided and used to form data generators for testing and training. The use of generators here is employed in order to be as memory-efficient as possible and load only the required subset of the dataset at a time.

In an effort to fully utilize the architecture described above, an effective flow of communication with the person (or a group of people) responsible for the

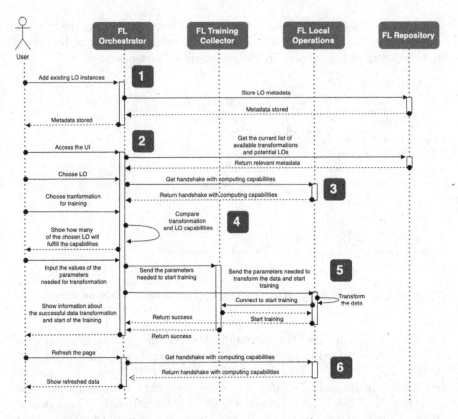

Fig. 4. A numbered version of the proposed data transformation communication scheme.

proper configuration of a given training process has been depicted in Fig. 4. This process should, first, ensure connection with required *FL Local Operations* and *FL Repository*. This exchange of messages can be understood as the operational and repetitive phase of user involvement. The part of the configuration of the FL system that would be initial and largely unchanging, throughout the functioning of the system, will be separately determined by the owners' consortium, at the start of the system's functioning. This will include information like the existence of *FL Local Operations* providing only inference in the system, the types of libraries that are to be used for training, or the initial quality of the data. The repetitive involvement, on the other hand, will define the variables that may change between separate training processes, like: which *Local Operations* instances will participate in a given training, how many of them, what kind of model should be used for this particular experiments, with what kind of configuration, etc.

The repetitive part of the configuration proceeds as follows (see Fig. 4). Aiming at providing a human-friendly interface, a UI connects the User with the *FL Orchestrator*, which allows to register and unregister the instances, from the list of Local Operations available for FL training. The information about these, potentially available, instances is the stored in the *FL Repository* (1).

As the UI is starting on the side of the User, the the *FL Orchestrator* downloads both the list of potential data transformations, and the *FL Local Operations* instances (2). Information concerning the system requirements, needed to run data transformation, selected by the User, along with a set of parameters needed as input of the transformation and a short description of what exactly is the transformation's purpose are also included in this list.

Subsequently, the *FL Orchestrator* checks the factual availability of the *FL Local Operations* instances, by sending them a specific HTTP request. In the response from *Local Operations*, the *FL Orchestrator* receives a list of the current computing capabilities of the instance, containing information about the amount of free RAM and storage, the existence of GPU on the device, as well as the preinstalled libraries and their versions (3). The *FL Orchestrator* is then periodically (for example on each refresh of the UI triggered by the user) supposed to resend these messages and ensure that the list of capabilities still is consistent with reality (6).

With this information, the *FL Orchestrator* determines whether a given *FL Local Operation* instances will be able to apply a selected data transformation. The *FL Orchestrator* can check if the *FL Local Operations* instance fulfills or even exceeds the hardware requirements and contains a compatible version of all of the libraries needed to run a given transformation (4). Upon discovery of situation that would preclude given *FL Local Operations* instance from participation in training, it will be removed from the list of FL participants. Next, *FL Orchestrator* parses obtained data to show the User the potential effects of choosing a given transformation. For instance, such decision may result in a substantial reduction of the number of *FL Local Operations* available to participate in training. However, the User makes the final choice of whether to proceed, or not. If the User places valid input parameters in the form, and triggers the training process to start, the right configuration is then transported to both the *FL Training Collector*, as well as to the appropriate *FL Local Operations*. The *FL Local Operations* should then be able to preprocess their local datasets and start training (5).

5.2   Privacy and Security

Since the scanners participating in the training in the proposed use case may have varying computing capabilities, belong to diverse organizations, and operate in different conditions, a solution that would enable the stakeholders to make an informed decision about the level of privacy in their FL system is needed. Additionally, the models employed for this specific use case tend to be of significant size (up to 180 MB in an uncompressed form), which indicates the need for higher computing capabilities in order to encrypt them using HE, which is quite computationally intensive even for smaller models [30]. In some cases, the inclusion of organizations with strict requirements regarding privacy throughout training, may necessitate in joint use of both HE and DP. In other cases, though, due to the hardware limitations of the scanners the additional computing capabilities necessary for the use of HE could be deemed unreasonable, while the degradation of final model performance would prevent the use of DP.

Based on this information, in the developed solution, we have decided to support hybrid approach combining HE and DP with adaptive gradient clipping on the local level, combined with random noise added on the global level. This choice is in line with our project paradigm providing maximal flexibility to the end user, which should then be able to choose whether they want to employ one of these approaches, none or both, and how rigorously. Regardless of the final deployment decision, either approach has to be (and is) supported. It is thus to the user can to decide how to balance sacrificing computing efficiency (HE) and model performance (DP).

To ensure the privacy and security of the resulting system, a couple of techniques have been employed. First, the User can define the privacy configuration of a given training job by filling out the correct form on the *FL Orchestrator*. Currently, User can then decide whether a training process will be conducted with the homomorphic encryption of the local updates, the adaptive clipping of the local updates and gaussian noise added to the global weight aggregation, none or both of these mechanisms. Moreover, the User can manipulate various DP input parameters, like the amount of noise added to the aggregated weights, and therefore configure to an extent the acceptable loss of model performance.

Since much of the communication in the resulting FL IoT system will involve not only exchanging private data but also notoriously insecure pickle files, SSL is used in order to provide end-to-end encryption for all exchanges.

6   Concluding Remarks

Running a FL-based system in a production for real-life use cases requires not only a distributed infrastructure with efficient communication between software elements and accurate ML models. What needs to be considered, before a deployment of such a system, and before an execution of training jobs, are requirements coming from the characteristics of the environment, both technical and business-specific. The latter covers issues related to "stability" of the ecosystem, network restrictions, etc. The former includes requirements for a level of privacy and security to be preserved, as well as data preprocessing, which is an crucial step in real-life deployments done on non-artificial data sets. The described ASSIST-IoT FL Reference Architecture covers all crucial elements that should be deployed in a FL system. Moreover, it is designed to be ready to deal with any additional issues that may appear when moving experimentations outside of the simulation platforms/environments.

References

1. Chamikara, M.A., Bertok, P., Khalil, I., Liu, D., Camtepe, S.: Privacy preserving distributed machine learning with federated learning. Comput. Commun. **171**, 112–125 (2021)
2. Introducing Federated Learning into Internet of Things ecosystems - preliminary considerations, July 2022. https://doi.org/10.48550/arXiv.2207.07700
3. Bellet, A., Kermarrec, A., Lavoie, E.: D-cliques: compensating noniidness in decentralized federated learning with topology. CoRR arXiv:2104.07365 (2021)
4. Beutel, D.J., Topal, T., Mathur, A., Qiu, X., Parcollet, T., Lane, N.D.: Flower: a friendly federated learning research framework. CoRR arXiv:2007.14390 (2020)
5. Bhowmick, A., Duchi, J.C., Freudiger, J., Kapoor, G., Rogers, R.M.: Protection against reconstruction and its applications in private federated learning. ArXiv arXiv:1812.00984 (2018)
6. Bilal, M., Ali, G., Iqbal, M.W., Anwar, M., Malik, M.S.A., Kadir, R.A.: Auto-prep: efficient and automated data preprocessing pipeline. IEEE Access **10**, 107764–107784 (2022)
7. Bonawitz, K.A., et al.: Practical secure aggregation for federated learning on user-held data. In: NIPS Workshop on Private Multi-Party Machine Learning (2016). arXiv:1611.04482
8. Byrd, D., Polychroniadou, A.: Differentially private secure multi-party computation for federated learning in financial applications. In: Proceedings of the First ACM International Conference on AI in Finance. ICAIF 2020, Association for Computing Machinery, New York, NY, USA (2021). https://doi.org/10.1145/3383455.3422562
9. Cao, J., Zhang, Q., Shi, W.: In Edge Computing, pp. 59–70. Springer International Publishing, Cham (2018)
10. Chabanne, H., de Wargny, A., Milgram, J., Morel, C., Prouff, E.: Privacy-preserving classification on deep neural network. IACR Cryptol. ePrint Arch. **2017**, 35 (2017)
11. Chen, Z., Li, D., Zhu, J., Zhang, S.: DACFL: dynamic average consensus based federated learning in decentralized topology (2021)
12. Eichner, H., Koren, T., McMahan, H.B., Srebro, N., Talwar, K.: Semi-cyclic stochastic gradient descent. CoRR arXiv:1904.10120 (2019)
13. Eschweiler, D., Spina, T.V., Choudhury, R., Meyerowitz, E.M., Cunha, A., Stegmaier, J.: CNN-based preprocessing to optimize watershed-based cell segmentation in 3d confocal microscopy images. In: 2019 IEEE 16th International Symposium on Biomedical Imaging (ISBI 2019), pp. 223–227 (2019)
14. Fahl, S., Harbach, M., Perl, H., Koetter, M., Smith, M.: Rethinking SSL development in an appified world. In: Proceedings of the 2013 ACM SIGSAC Conference on Computer & Communications Security, pp. 49–60. CCS 2013, Association for Computing Machinery, New York, NY, USA (2013). https://doi.org/10.1145/2508859.2516655
15. Hong, C.H., Varghese, B.: Resource management in fog/edge computing: a survey on architectures, infrastructure, and algorithms. ACM Comput. Surv. **52**(5), 1–37 (2019)
16. Hu, Q., Sun, P., Yan, S., Wen, Y., Zhang, T.: Characterization and prediction of deep learning workloads in large-scale GPU datacenters. CoRR arXiv:2109.01313 (2021)

17. Jiang, J., Hu, L., Hu, C., Liu, J., Wang, Z.: BACombo-bandwidth-aware decentralized federated learning. Electronics. **9**(3), 440 (2020). https://www.mdpi.com/2079-9292/9/3/440, https://doi.org/10.3390/electronics9030440

18. Jiang, L., Tan, R., Lou, X., Lin, G.: On lightweight privacy-preserving collaborative learning for internet of things by independent random projections. ACM Trans. Internet of Things **2**, 1–32 (2021)

19. Khan, L.U., Saad, W., Han, Z., Hossain, E., Hong, C.S.: Federated learning for internet of things: recent advances, taxonomy, and open challenges. CoRR arXiv:2009.13012 (2020)

20. Kumar, S., Schlegel, R., Rosnes, E., Amat, A.G.: Coding for straggler mitigation in federated learning (2021)

21. Lee, J., Oh, J., Lim, S., Yun, S., Lee, J.: Tornadoaggregate: accurate and scalable federated learning via the ring-based architecture. CoRR arXiv:2012.03214 (2020)

22. Li, L., Ono, K., Ngan, C.K.: A deep learning and transfer learning approach for vehicle damage detection. In: The International FLAIRS Conference Proceedings, vol. 34 (2021). https://doi.org/10.32473/flairs.v34i1.128473

23. Lu, Y., Huang, X., Dai, Y., Maharjan, S., Zhang, Y.: Blockchain and federated learning for privacy-preserved data sharing in industrial IoT. IEEE Trans. Indus. Inform. **16**(6), 4177–4186 (2020). https://doi.org/10.1109/TII.2019.2942190

24. Martynova, M., Kaas, O.: A novel methods based on clustering algorithms as the neural network preprocessing, pp. 317–322 (2019). https://doi.org/10.1109/SAMI.2019.8782767

25. McMahan, H.B., Moore, E., Ramage, D., Arcas, B.A.: Federated learning of deep networks using model averaging. CoRR arXiv:1602.05629 (2016)

26. McMahan, H.B., Ramage, D., Talwar, K., Zhang, L.: Learning differentially private language models without losing accuracy. CoRR arXiv:1710.06963 (2017)

27. Mhaisen, N., Abdellatif, A.A., Mohamed, A., Erbad, A., Guizani, M.: Optimal user-edge assignment in hierarchical federated learning based on statistical properties and network topology constraints. IEEE Trans. Netw. Sci. Eng. **9**(1), 55–66 (2022). https://doi.org/10.1109/TNSE.2021.3053588

28. Mothukuri, V., Parizi, R.M., Pouriyeh, S., Huang, Y., Dehghantanha, A., Srivastava, G.: A survey on security and privacy of federated learning. Fut. Gener. Comput. Syst. **115**, 619–640 (2021). https://doi.org/10.1016/j.future.2020.10.007, https://www.sciencedirect.com/science/article/pii/S0167739X20329848

29. Nguyen, D.C., Ding, M., Pathirana, P.N., Seneviratne, A., Li, J., Vincent Poor, H.: Federated learning for internet of things: a comprehensive survey. IEEE Commun. Surv. Tutor. **23**(3), 1622–1658 (2021)

30. Park, J., Lim, H.: Privacy-preserving federated learning using homomorphic encryption. Appl. Sci. **12**(2), 734 (2022)

31. Qiu, X.,et al.: A first look into the carbon footprint of federated learning. CoRR arXiv:2102.07627 (2021)

32. Roh, Y., Heo, G., Whang, S.E.: A survey on data collection for machine learning: a big data - AI integration perspective. IEEE Trans. Knowl. Data Eng. **33**(4), 1328–1347 (2021). https://doi.org/10.1109/TKDE.2019.2946162

33. Satapathy, A., Livingston, L.M.J.: A comprehensive survey on SSL/TLS and their vulnerabilities. Int. J. Comput. Appl. **153**, 31–38 (2016)

34. Singh, R., Ayyar, M.P., Sri Pavan, T.V., Gosain, S., Shah, R.R.: Automating car insurance claims using deep learning techniques. In: 2019 IEEE Fifth International Conference on Multimedia Big Data (BigMM), pp. 199–207 (2019). https://doi.org/10.1109/BigMM.2019.00-25

35. Song, M., et al.: Analyzing user-level privacy attack against federated learning. IEEE J. Sel. Areas Commun. **38**(10), 2430–2444 (2020). https://doi.org/10.1109/JSAC.2020.3000372

36. Tak, A., Cherkaoui, S.: Federated edge learning: design issues and challenges. IEEE Netw. **35**(2), 252–258 (2021). https://doi.org/10.1109/MNET.011.2000478

37. Thakkar, O., Andrew, G., McMahan, H.B.: Differentially private learning with adaptive clipping. CoRR arXiv:1905.03871 (2019)

38. Wang, S., McDermott, M.B.A., Chauhan, G., Ghassemi, M., Hughes, M.C., Naumann, T.: Mimic-extract: a data extraction, preprocessing, and representation pipeline for mimic-iii. In: Proceedings of the ACM Conference on Health, Inference, and Learning, pp. 222–235. CHIL 2020, Association for Computing Machinery, New York, NY, USA (2020). https://doi.org/10.1145/3368555.3384469

39. Yin, X., Zhu, Y., Hu, J.: A comprehensive survey of privacy-preserving federated learning: a taxonomy, review, and future directions. ACM Comput. Surv. **54**(6), 1–36 (2021)

40. Zheng, H., Hu, H., Han, Z.: Preserving user privacy for machine learning: local differential privacy or federated machine learning? IEEE Intell. Syst. **35**(4), 5–14 (2020). https://doi.org/10.1109/MIS.2020.3010335

41. Zhu, H., Xu, J., Liu, S., Jin, Y.: Federated learning on non-IID data: a survey. CoRR arXiv:2106.06843 (2021)

42. Fornes-Leal, A., et al.: ASSIST-IoT: A Reference Architecture for Next Generation Internet of Things (2022). https://doi.org/10.3233/FAIA220243

Blockchain Based B-Health Prototype for Secure Healthcare Transactions

Puneet Goswami[1,2], Victor Hugo C. de Albuquerque[1], and Lakshita Aggarwal[2(✉)]

[1] Federal University of Ceará, Fortaleza, Brazil
goswamipuneet@gmail.com, victor.albuquerque@ieee.org
[2] SRM University, Delhi NCR, Sonepat, India
Lakshitaaggarwal31@gmail.com

Abstract. The rapid advances in crypto wallet are re-defining privacy around transactions. A Crypto wallet comprises of software containing private and public keys and uses Blockchain to send and receive currency. Since the wallet will be based on the Blockchain, every transaction will be recorded, and every transaction will be stored. The privacy concerns have emerged out interaction between the normal transactions being made and the transactions made using blockchain in many fields like healthcare, defence etc. For the check on transactional details even the currency is not stored at one location; instead, they all exist as a transaction records on the Blockchain. In creating a Crypto Digital Wallet, we will solve traditional wallet problems. As these wallets store private and public keys, a user is facilitated with various operations such as sending or receiving coins, Portfolio balance, and crypto currency trading. This also ensures the user's privacy by using a hexadecimal address of the wallet. However, the currency's address to be exchanged differs from one service provider to another. Therefore, a Blockchain-based wallet provides all the necessary features for safe and secure transfers and exchange of funds between different healthcare consumers. It will also allow easy exchange of currency between different countries when patients are treated in different countries the issue of currency exchange will also be conquered. The renowned papers from various journals are studied and then it is discovered that there is no crypto wallet that can provide personalized point of service to the patients. In this paper, we have tried creating a "Crypto wallet based on Blockchain," to manage healthcare data the issues of security and data breach will be eradicated.

Keywords: Cryptography · Blockchain · Web 3.0 · Crypto currency · Healthcare · Data analysis

1 Introduction

In today's world, there are many crypto wallets; most are worldwide platforms, with only a few developed in India. During our analysis of several crypto wallets produced in India and now in use, we discovered that many lack capabilities, such as being unavailable for the daily transaction and lacking statistics on the current crypto market. We examined and evaluated several platforms accessible in India, such as coin switch kuber. Still, none

© The Author(s), under exclusive license to Springer Nature Switzerland AG 2023
S. Sachdeva et al. (Eds.): BDA 2022, LNCS 13830, pp. 70–85, 2023.
https://doi.org/10.1007/978-3-031-28350-5_6

of them were adequate for competing with the foreign platform and meeting the needs of Indian crypto users. As we examined further, India will emerge as a top crypto exchanger due to rising demand. Therefore, we require a platform that can provide us with what the Indian community wants. We discovered that wallets lacked an exemplary user interface or were just showpieces that did nothing. However, competition among platforms is increasing globally, and it will soon be too late if we do not provide a platform. With the government's growing limitations on cryptocurrency and many large corporations pulling back from trading, now is the time for a Crypto Wallet that can meet the needs of both consumers and governments while emerging as a new worldwide competitor.

In the era of 21st century after entering the age of digital banking, we still follow some of the traditional E-wallet methods [3], and we face several security loopholes and transactional time specific to payment gateway. Technologies like blockchain, Cryptography and Web 3.0 helps in assisting the problem. Using cryptocurrency, a digital currency that will be stored in a crypto wallet. The crypto wallet has the functionality of storing every transaction of the amount. In this wallet, the money is converted and added as a form of coin, such as bit coin, ethereum, etc. The currency in this wallet is not stored at one location; instead, they all exist as transaction records on the blockchain [18]. As these wallets store private and public keys, a user is facilitated with various operations such as sending or receiving coins, monitoring coin balance, and trading the currencies on a portfolio using the wallet. This also ensures the user's privacy by using a hexadecimal (16 bit) address of the wallet. However, the currency's address to be exchanged differs from one service provider to another. The wallet-based software can be accessed from devices, including mobile and web, and the privacy and identity of the user are maintained.

1.1 Blockchain

Blockchain [4] has become one of the most creative application models because to its capacity to incorporate consensus procedures, distributed data storage [20], digital encryption technology, peer-to-peer transmission, and other computer technologies. It has given a secure and decentralized platform for information exchange.

The data layer, network layer, consensus layer, incentive layer, contract layer, and application layer make up the blockchain architecture [9].

- The necessary asymmetric encryption and timestamp technologies are contained in the data layer, which also contains the bottom layer data blocks.
- Code and transactions [19] that come in at a certain time are appended to a new block with a timestamp and connected to the main deterrent to be added to the chain as a new block.
- The incentive layer combines economic elements into the blockchain, such as the currency system and the money distribution mechanism, to reward bitcoin miners (who generate the next block).
- The contract layer network, which consists of code and algorithms, includes linked algorithms, smart contracts [13], and scripts. For flexible data programming and operation in a blockchain system, the contract layer is necessary.

- The application layer incorporates scenarios and use cases. The timestamp chain structure, distributed node consensus, and adaptable programmable intelligent contract
make the blockchain both technical and creative.

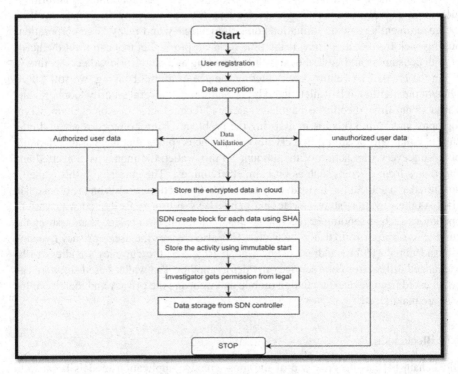

Fig. 1. Blockchain architecture

Figure 1 elaborates the procedure to be followed if the user is new he/she will do
the registration. The data will be encrypted to provide security. If the data gets validated
the user gets authorized to store the data on cloud. Software Defined Networking (SDN)
creates block for each data using Secure Hash Algorithms (SHA) [5]. The block stores
all the activity using immutable start. All the investigators after successful verification
gets the permission from legal authorities. Data gets stored from SDN controller and the
process goes on. If the user is an unauthorized user that means the validation of the data
is unsuccessful then the user is directly redirected to stop the flow.

1.2 Cryptography

Using codes, cryptography is a method for protecting data and communications such
that only those who need to comprehend and process it may do so. Unwanted access to
information is therefore avoided. In order to modify communications in ways that make
them challenging to decode, cryptography techniques are derived from mathematical

concepts and a collection of rule-based computations known as algorithms. These algorithms create cryptographic keys, enable digital signatures, confirm data privacy, enable web browsing, and safeguard private transactions.

Figure 2 elaborates how the data gets encrypted/decrypted using cipher text. Sender encrypts the plain text using encryption key and algorithm. The text gets converted to cipher text. If the interceptor tries to intrude it he/she won't be able to do so. Now, when the receiver receives it decrypts the text using decryption algorithm, converts decrypted text to plain text. This mechanism allows the data to get secured from interceptors.

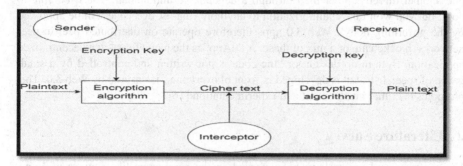

Fig. 2. Cryptography

There are three broader types of cryptography:

1. Symmetric Key Cryptography: Symmetric Key Systems have the disadvantage of needing the sender and receiver to securely exchange keys, despite being faster and easier to use. Data Encryption System is the most widely used symmetric key encryption method (DES). It is an encryption method in which the sender and recipient of a message use a single shared key to encrypt and decode communications.
2. Hash Functions: This algorithm employs no keys. The plain text is used to generate a hash value with a defined length, rendering it difficult to reverse-engineer the plain text's contents. Hash functions are used by several operating systems to safeguard passwords.

Asymmetric Key Cryptography: Using two keys, this technique encrypts and decrypts data. Data is encrypted with a public key and decrypted with a private key. Public and private keys don't necessarily mean the same thing. The intended recipient can only decode it because he has access to the private key, even if everyone else has access to the public key.

1.3 Web 3.0

A data-driven semantic web is a key component of web 3.0, a third-generation internet service for websites and apps that focuses on exploiting a machine-based [22] understanding of data. It incorporates every aspect of the internet that uses an individual's address to connect. The creation of more inventive, connected, and open websites is Web 3.0's ultimate objective.

Unlicensed and unreliable: Because Web 3.0 is built on open-source software, it is not only decentralised but also inconsistent because it lets users to communicate with one another directly rather than through a dependable middleman. You'll be able to do it, but you won't have authorization to anymore (that is, everyone will be approved by the governing body). Web 3.0 apps therefore operate on distributed peer-to-peer networks, blockchain, or a mix of these. A "dApp" is the name for such a decentralized application. Bottom-up design, or "the code is not written and controlled by a small group of specialists, but is developed in front of everyone," is ensured by Web 3.0. This promotes maximal involvement and experimentation [16].

2 Literature Survey

Various renowned papers have been studied on the parameters like support of coins, Portfolio management, Charts, exchange of coins, and storing of data. Various models [11, 14] like Meta mask, Coin switch, and trust wallet have been compared with our proposed model "BHealth" (Table 1).

Table 1. Comparison among traditional wallets and proposed solution

Parameter	Metamask	Coinswitch	Trust wallet	Proposed model: "BHealth"
Support of coins	×	✓	✓	✓
Portfolio management	×	✓	×	✓
Charts	×	✓	×	✓
Exchange of coins	✓	×	✓	✓
Storing	✓	×	✓	✓

The author discusses how Bitcoin-beyond blockchain work bridges those flaws and some unresolved problems. Cryptocurrencies Block chain's specifications and guarantees do not fit FinTech's requirements on security and privacy [24] from transaction throughput to primitives. It analyses the safeguarding process of the distributed database and suggests a solution for the challenges of retaining the information confidentiality in them without tokens based on Blockchain. The authors say that without mining and permits, Blockchain would significantly unravel the procedure to maintain the privacy [23] and validity of knowledge regarding bank transactions. In this work, Blockchain technology addresses the problem of cryptography consensus. And if there is a method to assure financial activity and transaction actions are stored in a particular database without the central authority's intervention. It analyses the leading design and technological features showcased by Blockchain and presents scenarios in which blockchain applications [12] can be used. The research paper focuses on using Blockchain as the Central Bank Digital Currency (CBDC) basic prototype technology. The Central Bank Digital [1] Currency prototype will benefit from the supervision, payment, and use of Blockchain technology. Problems such as safeguarding the confidentiality, transparency, and speed of user transactions should be resolved to use the Blockchain as CBDC's fundamental technology [15]. It is the first project that has been created with the collaboration of Blockchain, cryptography, and web 3.0. The research paper tells us about the Blockchain and cryptography used in real-world problems.

3   Problem Statement

The Tradition online wallets don't guarantee to customers that they are secured with the type of wallets they choose. No wallet holds cryptocurrency [6] for daily transactions in India (or many parts of the world), as many of the wallets are used to Purchase & Sell the cryptocurrency, like trading. No wallet has a suitable authentication protocol that secures the connection with a world-leading blockchain to verify a secure connection for the transaction. With time it has been proven that no such technology can empower the banking system [7] and protect customers' rights against financial fraud, as goodwill comes with a cost that is a trust which takes years to build.

3.1   Approach to Problem in Terms of Technology

In this evolving time, there are many traditional wallets for transaction and storing. But there is no guarantee of the security [2], and there are many loopholes in transaction (Table 2).

Table 2. Qualitative comparison in traditional wallets and BHealth CryptoWallet

Issues in traditional wallet	Proposed solution (BHealth CryptoWallet based on Blockchain)
Secured login facility gets hampered easily	Authentication by • Metamask • Infura
Password security not available	Hexadecimal bit password security
Transaction record security	Daily based transaction record security
Time consumption transactions	Transactions made in few time irrespective of currency exchange
Intermediate between customer to customer (C to C)	Person to Person (P to P)

By creating a "Crypto wallet based on Blockchain" the issues [10] of security and loopholes will be over.

Platforms to be used for creation of the Blockchain Based [17] app.

- Front end Web development
- Node.js (Back-end, runtime)
- JavaScript (Front-end)
- React (JavaScript library)
- Styled Component (styling)

Back end of Web development

- Sanity (Database)
- GROQ (Query fetcher)

Token

- Third web

Authentication

- Metamask
- Infura

Package Manager

- Npm
- Yarn

3.2 Requirements

See Table 3.

Table 3. Functional, Non-functional and logical requirements for proposed methodology

Functional requirements	Non-Functional requirements	Logical database requirements
The software system will do: • Portfolio management • Balance • Send crypto • Buy crypto • Sell crypto • Receive crypto • Staking	The software system provides: • Security • Maintainability • Portability • Reliability • Reusability • Flexibility • Scalability • Performance	System will be able to do: • Quick access (within ms) • Correct information • Server working

3.3 Proposed Methodology

The main objective of creating a Crypto Digital Wallet will solve the problems of traditional wallets. A Crypto wallet comprises software containing private and public keys and uses Blockchain to send and receive currency. Since the wallet will be based on the Blockchain, every transaction will be recorded, and every transaction will be stored. The transaction details or even currency are not stored at one location. Instead, they all exist as transaction records on the Blockchain. As these wallets store private and public keys, a user is facilitated with various operations such as sending or receiving coins, monitoring coin balance, trade the cash on a portfolio using the wallet. This also ensures the user's privacy by using a hexadecimal address of the wallet. However, the currency's address to be exchanged differs from one service provider to another.

There will be no interference from the government as there is no central bank in between. The transaction will be between person to person (P2P). So, the transaction will be faster and more secure.

Some parameters which will help to secure:

a) The wallet will be created using a phrase instead of a password, OTP, etc. This Phrase is provided to the specific user at the time of wallet creation.
b) Email must be used to verify and monitor the transaction activity. Each transaction using the wallet should be recorded on the Blockchain.
c) A customer can initiate the transaction using a wallet address for receiving and sending the transactions. It will also hide the customers' information from the outside world.

The earning method for us will be the transaction fee. We will receive a certain percentage of the amount transacting for every transaction.

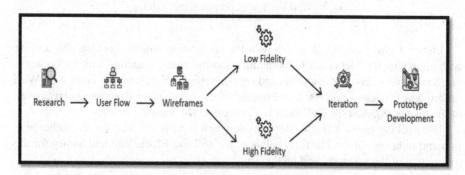

Fig. 3. Prototype development

It is the process (Fig. 3) in which we created the "BHealth" since the project should work in a flow to make a fast and stable process—starting with the research on the project with different IEEE certified papers—then creating a user flow of the project with modules and wireframes. A working prototype was made with the various combinations of the modules.

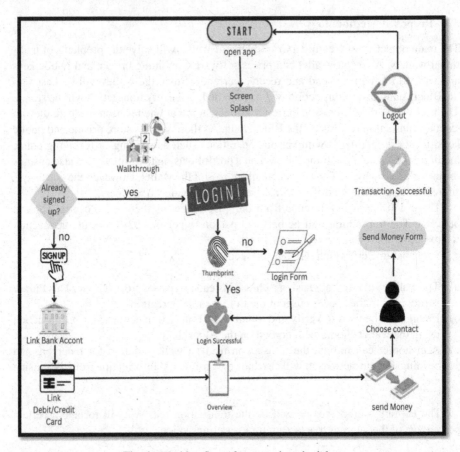

Fig. 4. Working flow of proposed methodology

Figure 4 illustrates the flow in which the application will be working and starting with login with the Metamask for authentication and then connecting with the backend, i.e., Sanity for receiving the queries and securing the database with the Third web. With the ongoing process, it will fetch the Portfolio balance and then for sending and receiving of the tokens [8]. And the end, there is a process of the logout.

For working of the BHealth, Metamask is first used in the base for the authentication and to bring out the Portfolio balance. We will use Blockchain and Sanity for the transaction of the token created on the Third web (Fig. 5).

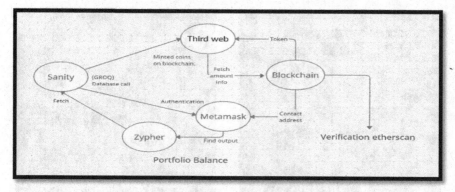

Fig. 5. BHealth Flow Process

Algorithm Snippets (Module-Wise)

```
const Portfolio = ({ twTokens, sanityTokens, walletAddress }) => {
  const [walletBalance, setWalletBalance] = useState(0)
  const [sender] = useState(walletAddress)

  const getBalance = async activeTwToken => {
    const balance = await activeTwToken.balanceOf(sender)

    return parseInt(balance.displayValue)
  }

  useEffect(() => {
    const calculateTotalBalance = async () => {
      setWalletBalance(0)

      sanityTokens.map(async token => {
        const currentTwToken = twTokens.filter(
          twToken => twToken.address === token.contractAddress,
        )

        const balance = await getBalance(currentTwToken[0])
        setWalletBalance(prevState => prevState + balance * token.usdPrice)
      })
    }

    if (sanityTokens.length > 0 && twTokens.length > 0) {
      calculateTotalBalance()
    }
  }, [twTokens, sanityTokens])
```

```
  useEffect(() => {
    twTokens.map(token => {
      if (token.address === selectedToken.contractAddress) {
        setActiveTwToken(token)
      }
    })
  }, [twTokens, selectedToken])

  useEffect(() => {
    const getBalance = async () => {
      const balance = await activeTwToken.balanceOf(sender)
      setBalance(balance.displayValue)
    }

    if (activeTwToken) {
      getBalance()
    }
  }, [activeTwToken])

  useEffect(() => {
    const url = builder.image(selectedToken.logo.asset._ref).url()
    setImageUrl(url)
  }, [selectedToken, builder])

  const sendCrypto = async () => {
    console.log('sending crypto')

    if (activeTwToken && recipient) {
      setAction('transferring')
      const result = await activeTwToken.transfer(
        recipient,
        amount.toString().concat('000000000000000000'),
```

Fig. 6. Portfolio module   **Fig. 7.** Transfer module

Figure 6, 7, 8 and 9 describes the algorithms used to develop the project i.e. portfolio module (describing how the blockchain modules and tokens interacts with each other for estimating balance and other crypto currency values of a token), transfer module (describes the details of sell or buy assets of the crypto currency), receive module (will show the updated transactions of the assets) and Dashboard module (shows the details of the transactions made).

Fig. 8. Receive module   **Fig. 9.** Dashboard module

3.4 Implementation Details and Issues

1. Understand Blockchain and Crypto
2. Use Standard Cryptocurrency Open-source Libraries

   a) Free libraries
   b) Supports Python, Java, Ruby

3. Use APIs

   a) Feature rich
   b) Easy synchronize with blockchain ecosystem

4. Select the Right Technology Stack

   a) Node.js (Back-end, runtime)
   b) JavaScript (Front-end)
   c) React (JavaScript library)
   d) Sanity (Database)
   e) GROQ (Query fetcher)
   f) Thirdweb

5. Accentuate Security

   a) Phrase login
   b) Connection Meta Mask

6. Database Connection
7. Features Addition

During the implementation of the BHEALTH there where many issues like:

a) Fetching of the data from the thirdweb.
b) Fetching the data from the databases.
c) Providing of the contract address.
d) Runtime error
e) Connection with database

4 Results and Conclusions

The advancement we provided in our crypto wallet with all the functionalities and all the accessibility in the security of sharing their keys, or sending or receiving of the crypto currency, providing the chart of the total portfolio, and the most important, we created the portfolio which calculates all the price of the crypto currency and summing it up, thus providing the total balance. The cost involved in treating the patients is easily exchanged using crypto wallets. The costs used in health information exchange provides ease of managing transactional data involved during the patient's outlook. The wallet can also store digital assets such as NFTs, which you may want to buy, sell, or trade. It's no surprise that we made security the most critical factor because of the intricacy of digital money. Cost, privacy and anonymity, usability, customer service, and features were other important considerations for a software wallet. While creation of the BHealth (Crypto Wallet) we analysed the functionality we are providing in the application (Table 4).

Table 4. Features of proposed application

Security	Cost and fees	Features	Privacy and anonymity
• Account access method • Transaction authorization method • Recovery method • Reputation • Hierarchical deterministic • Open-source code	• Device/account (one-time) fee • Send (recurring) fee • Receive (recurring) fee • Can customize fee	• Number of currencies supported • Fiat funding available • Frequency of update • Earn interest on crypto holdings • Portfolio balance • Staking option • Can buy/Sell crypto • Can swap/exchange/convert crypto-to-crypto	• Transaction anonymization method • Conjoin-enabled • New address generated

With the advancement in the world of technology, blockchain is working in a way that will be transparent to the world, which will have every information securely; it will also provide us the data as fast as possible with the security that is very tough to crack.

With this advancement, it's high time there is a wallet created in India for the consumers of India, which will be a transparent wall between the world government and its people. With the traditional method of transacting and storing the currency, the people can also use the secure way of transacting their money in the "Blocks of Chain," the secured and fastest way.

5 Future Work

The Cloud [21] wallet is in the early stage of its full development, in which we provided many unique features and functionality. With time improvements can be done as the wiser method for double factor authentication. The system would be able to support more crypto currency for transactions. The current lacks the mechanism for monitoring everyday transactions in real time. More such features will be improvised in the coming time as per the need. The BHealth will help in preserving data confidentiality and data breach. Such systems enhance the use case not only in health but also in military, defence, businesses etc.

Appendix

It includes the list of Graphical User Interface (GUI) figures obtained in the project for which the algorithms are already stated above. The below figures are given for reference (Figs. 10, 11 and 12).

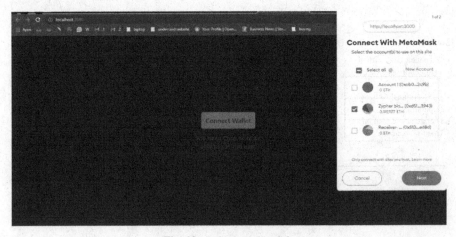

Fig. 10. Index page module

Fig. 11. Transfer module

Fig. 12. Asset module dashboard

References

1. Nakamoto, S.: Bitcoin: A Peer-to-Peer Electronic Cash System (2008). https://bitcoin.org/bit coin.pdf
2. Yadav, N.S.: Crypto wallet: a perfect combination with blockchain and security solution for banking. Int. J. Psychosoc. Rehabil. **24**(02), 6056–6066 (2020)
3. Jokić, S., Cvetković, A.S.: Comparative analysis of cryptocurrency wallets vs traditional wallets. Ekonomika **65**(3), 65–75 (2019)
4. Rezaeighaleh, H.: Improving Security of Crypto Wallets in Blockchain Technologies (2018)

5. Lin, I.-C., Liao, T.-C.: A survey of blockchain security issues and challenges. Int. J. Netw. Security **19**(5), 653–659 (2017)
6. Aggarwal, L., Singh, P., Singh, R., Kharb, L.: IoIT (Internet of Intelligent Things). In: Kaur, G., Tomar, P., Tanque, M. (eds.) Integrating Artificial Intelligence with IoT to Solve Pervasive IoT Issues, pp. 251–267. Elsevier (2021)
7. Rao, P., Bhasin, N., Goswami, P., Aggarwal, L.: Crypto currency portfolio allocation Using Machine Learning. In: 2021 3rd International Conference on Advances in Computing, Communication Control and Networking (ICAC3N), pp. 1522–1527 (2021). https://doi.org/10.1109/ICAC3N53548.2021.9725704
8. Eyal, I.: Blockchain Technology: Transforming Libertarian Cryptocurrency Dreams to Finance and Banking Realities Computer MDPI AG (2017)
9. Popova, N.A., Butakova, N.G.: Research of a possibility of using blockchain technology without tokens to protect banking transactions. In: Proceedings of the 2019 IEEE Institute of Electrical and Electronics Engineers (2019)
10. Zheng, Z., Xie, S., Dai, H., et al.: An overview of blockchain technology: architecture, consensus, and future trends. In: IEEE 6th International Congress on Big Data (2017)
11. Joshi, A.P., Han, M., Wang, Y.: A survey on security and privacy issues of blockchain technology. Math. Found. Comput. **1**(2), 121 (2018)
12. Kharb, L., Aggarwal, L., Chahal, D.: A contingent exploration on big data tools. In: Bindhu, V., Chen, J., Tavares, J.M.R.S. (eds.) International Conference on Communication, Computing and Electronics Systems. LNEE, vol. 637, pp. 743–753. Springer, Singapore (2020). https://doi.org/10.1007/978-981-15-2612-1_71
13. Kosba, A., Miller, A., Shi, E., Wen, Z., Papamanthou, C.: Hawk: the blockchain model of cryptography and privacy-preserving smart contracts. In: 2016 IEEE Symposium on Security and Privacy (2016)
14. Chen, W., Xu, Z., Shi, S., et al.: A survey of blockchain applications in different domains. In: International Conference on Blockchain Technology and Applications (ICBTA) (2018)
15. Buterin, V.: A Next-Generation Smart Contract and Decentralized Application Platform (2013)
16. Coinbase home page. https://www.coinbase.com/
17. Pawar, V., Patel, A.K., Sachdeva, S.: Blockchain-enabled system for interoperable healthcare. In: Sachdeva, S., Watanobe, Y., Bhalla, S. (eds.) Big-Data-Analytics in Astronomy, Science, and Engineering. LNCS, vol. 13167. Springer, Cham (2022). https://doi.org/10.1007/978-3-030-96600-3_9
18. Pawar, V., Sachdeva, S.: CovidBChain: framework for access-control, authentication, and integrity of Covid-19 data. Concurr. Comput. Pract. Exp. **34**(28), e7397 (2022). https://doi.org/10.1002/cpe.7397
19. Puri, V., Kaur, P., Sachdeva, S.: (k, m, t)-anonymity: enhanced privacy for transactional data. Concurr. Comput. Pract. Exp. **34**(18), e7020 (2022). https://doi.org/10.1002/cpe.7020
20. Aggarwal, L., Chahal, D., Kharb, L.: Pruning deficiency of big data analytics using cognitive computing. In: 2020 International Conference on Emerging Trends in Communication, Control and Computing (ICONC3), pp. 1–4 (2020). https://doi.org/10.1109/ICONC345789.2020.9117504
21. Madan, S., Goswami, P.: A privacy preserving scheme for big data publishing in the cloud using k-anonymization and hybridized optimization algorithm. In: 2018 International Conference on Circuits and Systems in Digital Enterprise Technology (ICCSDET), pp. 1–7 (2018). https://doi.org/10.1109/ICCSDET.2018.8821140
22. Rahul, K., Banyal, R.K., Goswami, P., Kumar, V.: Machine learning algorithms for big data analytics. In: Singh, V., Asari, V.K., Kumar, S., Patel, R.B. (eds.) Computational Methods and Data Engineering. AISC, vol. 1227, pp. 359–367. Springer, Singapore (2021). https://doi.org/10.1007/978-981-15-6876-3_27

23. Madan, S., Goswami, P.: A novel technique for privacy preservation using k-anonymization and nature inspired optimization algorithms. In: Proceedings of International Conference on Sustainable Computing in Science, Technology and Management (SUSCOM), Amity University Rajasthan, Jaipur - India, 26–28 February 2019 (2019). https://ssrn.com/abstract=3357276. http://dx.doi.org/10.2139/ssrn.3357276

24. Madan, S., Goswami, P.: Recent Advances in Computer Science and Communications (Formerly: Recent Patents on Computer Science). Adaptive Privacy Preservation Approach for Big Data Publishing in Cloud using k-anonymization, vol. 14, no. 8, pp. 2678–2688 (2021). https://doi.org/10.2174/2666255813999200630114256

A Blockchain-Based Approach for Audit Management of Electronic Health Records

Rashmi P. Sarode$^{(\boxtimes)}$, Yutaka Watanobe, and Subhash Bhalla

Department of Information Systems, University of Aizu, Aizuwakamatsu, Japan
rashmipsarode@gmail.com, yutaka@u-aizu.ac.jp

Abstract. The maintenance of proper health records is essential to patient health care. Electronic Health Records (EHR) are now replacing traditional Manual Health Records. Audit logs or Audit Trails are a record of events and changes done in a system. Majority of hospitals are required to maintain an audit trail of each and every EHR. Currently, the audit trail is stored in relational databases, which can be easily modified and trust can be lost in the process. Also, third-party audit trails are inefficient, costly, and time-consuming. Replication on Blockchain would be a viable method for securing audit trails so that they are secure, transparent, and immutable without the need for third-party intervention. In this manuscript, we have proposed an Audit Management System where immutable audit trail of EHR can be generated on Blockchain. In this system, a physician or other medical authority having access to this audit trail can easily verify all the consultations, procedures and prescriptions given to the patient in a chronological manner.

Keywords: Electronic health records · Electronic medical records · Audit trail · Audit in blockchain

1 Introduction

A handwritten system was used to keep track of health information in the healthcare field. Manual medical recording systems were sluggish, unsecured, and lacked proper organization. An EHR (Electronic Health Record) and EMR (Electronic Medical Record) system provide better security and ease of access while attempting to provide appropriate access controls [1].

EMRs are medical data in the form of digital records that is easy to store, update, and exchange between healthcare organizations anywhere and anytime. EHR refers to a repository of patient data in digital form, maintained and shared securely, accessible by numerous authorized users [2]. The EHR is classified as an inter-organizational system, whereas the EMR is typically regarded as an intra-organizational system [3]. Patients' EMRs and EHRs are distinct from each other. It is possible for medical professionals, patients, and anyone else with permission, to access electronic medical records accurately and promptly [4].

Some EHRs include applications that provide audit trails for auditing and system transparency. These applications continuously monitor the database of the application and generate an audit trail record whenever an object's value changes. Due to the client-server interaction, however, audit trails are susceptible to a single point of failure, allowing an adversary to externally and internally change the database and audit trails. Replicating audit trails across all applications is an apparent method for protecting them against single points of failure. This will increase the cost of an attack for the adversary, as altering audit trails requires an attack on all applications. This replication of audit trails can be accomplished employing blockchain technology to offer safe, transparent, and immutable audit trail management without the requirement for a trusted middleman [5].

In addition, an adversarial party attempting to hack the system will be necessary to replace all logs maintained by each peer. This, in turn, increases the cost and complexity of the attack, thus enhancing the audit log application's total defensive capability [6]. With this in mind, we embark on an effort to build an audit trail using blockchain technology. The goal of this paper is to use blockchain technology as a solution to the problem of distributed data in hospitals and generate an audit trail of consultations and investigations that can be made without making significant changes to the existing hospital database.

The rest of the manuscript is as follows: Sect. 2 consists of Related Works, Sect. 3 describes the Proposed System in detail, Sect. 4 presents the Benefits of the Proposed Architecture, Sect. 5 elaborates on the Challenges and Discussion, and finally, Sect. 6 concludes the manuscript with Summary and Conclusions.

2   Related Works

Numerous researchers have conducted extensive research on Blockchain and Smart Contracts in the Medical field; however, the research in Audit Management is limited.

Kaushik et al. (2022) [7] investigate the potential advantages and applications of blockchain and quantum technology in the fields of medicine, pharmacy, and healthcare systems. The application of blockchain technology in the fight against the COVID-19 outbreak, as well as in other healthcare systems of a similar nature, has proven to be quite beneficial. Modern medical care systems feature sensor data that can be utilised to monitor patients while safeguarding their privacy and medical records' confidentiality. Using technologies like as quantum computing and blockchain could make it possible to manage patient data more rapidly and securely. Both blockchain and quantum technologies have been the topic of considerable speculation. This investigation attempts to determine the existence of operational medical uses for either of these technologies in the present scenario.

Sharma, et al. (2021) [8] proposed a multi-constraint and multi-objective simulation-optimization strategy for scheduling linear and nonlinear dynamic and controlled drone movement models. The current COVID-19 pandemic situation required the development of tools for the precise and rapid distribution of disease, impacts, causes, and treatment-related data. The performance of both types of drone-based smart healthcare systems can be evaluated based on their ability to maintain a stable attitude for both inside and outdoor activities. This effort began with an introduction to COVID-19, identification of hotspots, monitoring, and effects on social and economic sectors. This work examines drone-based COVID-19 interior and outdoor operations, including sanitization, monitoring, data collecting, and sharing. In results and analysis, the number of patients admitted, the number of drones utilised, and the proportion of individual drone utilisation are reviewed and discussed.

A proposed solution in healthcare by Kumar, et al. (2020) [9] utilises blockchain as a safeguard against the failure of central authority and employs a decentralised strategy. Most of the smart healthcare systems prioritise either data security issues or huge data management. The authors propose a simulation-optimization strategy to enhance the overall system and subsystem performance. The incorporation of blockchain technology into the healthcare system gives a range of characteristics, including automation, transparency, and security, to applications such as healthcare.

BlockTrail, a multilayer blockchain system was proposed by Ahmad, et al. (2019) [5] that utilizes the hierarchical distribution of copies in audit trail applications to minimize system complexity and boost performance. BlockTrail divides a single ledger into many chains managed at different system tiers. The prototype of BlockTrail is based on an audit trail application and uses the PBFT protocol to enhance consensus among copies. Experiments indicate that, compared to typical blockchains, BlockTrail is more efficient, with manageable delays.

Ahmad, et al. (2018) [6] proposed BlockAudit, a blockchain-based audit log system that uses blockchain technology's security capabilities to produce distributed, append-only, and tamper-proof audit logs. The authors put their design into practice using Hyperledger, then assessed the system's performance in terms of latency, network size, and payload size. Three assessment metrics are used to analyze the system's performance: latency, network size, and payload size.

3   Proposed System

We propose a blockchain-based audit management system in which a physician or medical authority with access to a patient's EMR can chronologically verify all consultations, procedures, and prescriptions issued to the patient.

3.1   System Architecture

Consider the case of three hospitals. The hospitals store data on a web-based UI (User Interface) that stores the data on a shared database. A summary of the data collected in the hospital databases is stored on Blockchain so that there is a note in brief about every visit and investigation carried out concerning the patient in a time-stamped manner. The hospitals register on Blockchain through Smart Contract. This is depicted in Fig. 1.

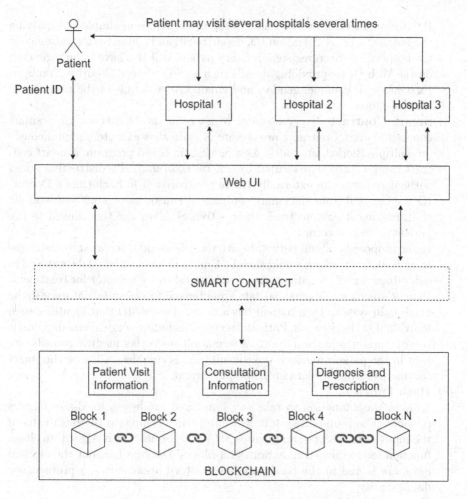

Fig. 1. Proposed System Architecture.

The challenge here is to create an immutable and sequential record of (i) the visits of a patient to these hospitals, (ii) investigations and procedures carried out, (iii) inferences made, and (iv) treatments and medicines prescribed. The steps for the Proposed System process are as follows:

(a) **Registration**: Each hospital has to be registered on the Web UI by providing a unique registration id, a name, an address, and an email address. These facts are verified by medical associations and other regulatory agencies before being recorded in the blockchain and utilized as the unique Hospital-

ID. Adding medical professionals to the blockchain is as simple as supplying the hospital's ID, registration ID, department, and contact information once the hospital has been registered. Every patient will also have to be registered on the Web UI by providing details such as SSN (Social Security Number), First name, last name, gender, and email. Other details of the patient can also be added.

(b) **Smart Contract**: Every hospital registers in the blockchain via a smart contract. Smart Contracts are scripts of code that execute simultaneously on multiple Blockchain nodes. As a blockchain-based program, a smart contract can be correctly executed by a network of mutually distrusting nodes without requiring an externally trusted authority [10]. Each time a Doctor-ID is generated from the smart contract, it can be associated with specific staff, making it easy to track their activities. They are then linked to the patient's medical record.

In the proposed system, patient health records would be created by the registration contract, which would then be kept in blocks of hashed data stored in a database server. A patient goes to a hospital to see a doctor for treatment. Patient's medical and personal data are entered into a Hospital database in the blockchain system; each patient has a unique Patient-ID that identifies each individual in the network. Patients' records, including surgical and diagnostic records, prescription and dosage information, and other medical records, are kept in the patient file once the consultation is complete. Finally, the smart contract receives the data in the proper format.

(c) **Hash Function**:
Using a Hash function to take any input and produce a fixed-size hash is possible. The proposed system will automatically update the hash value if the input data changes, and it uses SHA-256 (Secure Hash Algorithm) hash function for hashing transactions and blocks. Data hashes, not the original data, are stored in the blockchain's repository because it is a protocol for data security.

(d) **Transaction Management**:
The proposed system's smart contract is constructed in such a way that it must generate an n-count number of health blocks. Each block is made specifically for a particular transaction, in this case, every patient's visit to the hospital (consultation). As soon as a patient's medical records are entered into a blockchain, the patient can access their data. With a unique patient-ID, the system automatically accesses prior patient health records. No one can alter the data on a blockchain since it is immutable. Preventing and detecting alterations to a patient's medical record is made easier with the proposed approach, which is very useful while performing audits. A group of pre-approved participants, including the patient, doctor, and an authorized hospital, will be responsible for making necessary changes to the system.

A patient may visit several hospitals and diagnostic centers during his routine, and if all of these are registered on the blockchain, then the entire trail will be available. The patient can access this data via a web UI where he can provide his ID and authentication information to retrieve data.

Further details can be obtained from the respective hospital's databases. However, a complete trail will be available to him so that he does not miss any vital information. Care must be taken to ensure that only one Patient ID is used in all hospitals and diagnostic centres that the patient visits. This ID is provided by the blockchain smart contract used to register the patient. A patient should be registered only once on the blockchain. This can be enforced by using a unique ID such as the patient's social security number while registering.

3.2 Proposed System Implementation

The proposed system is implemented using Ethereum Blockchain Ganache [11] and Truffle [12] that is used to upload smart contracts in Blockchain. The smart contract is critical to implementing the blockchain because it allows the various stakeholders in the system to carry out their agreements.

For auditing, we focus on adding and retrieving hashes on the blockchain in our proposed system. Patients, doctors, and other healthcare professionals can access the blockchain hash. As specified in the registration contract, anyone attempting to do an audit must have permission and approval from the patient being audited. The system sends a false statement if the doctor is not permitted, and the session is terminated. Smart contracts check the patient's and doctor's addresses against information already stored on the blockchain.

We have created the smart contract for Patient Registration and validated it in Blockchain. The patient data is entered into the Web UI, which exists in the hospital. The Node.js module posts the relevant data in the smart contract, which is executed. The code for the createPatient() function in the smart contract is shown in Fig. 2. The data is then stored in the Blockchain via a smart contract transaction which is then validated in Blockchain as seen in Fig. 3. The other smart contracts can be created using a similar method.

4 Benefits of Proposed Architecture

The benefits of such an architecture are that the hospitals can choose any proprietary or web based system to suit their requirements. In addition, a gateway is provided to access the entire history of the patient's health records and investigations. Suppose the patient goes to any other hospital (Hospital 2). In that case, the same Patient-ID could be referred, for example, in situations where a patient would need a second opinion from a different doctor due to relocation or other reasons. Doctors in this situation can review the patient's previous treatment logs to plan subsequent treatment. For further treatment, doctors can access patient records in the blockchain with the patient's permission (the patient provides his unique Patient-ID), and patients can share their health data with any doctor in the network.

In the proposed approach, the patient's medical records are safeguarded. The system allows the registered user (patient) and any authorized health service provider in the blockchain network (in case of emergency) to access and

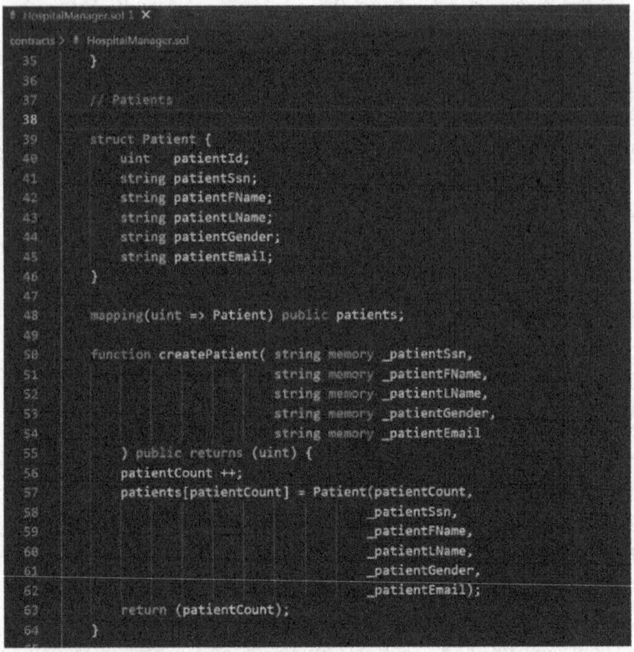

```
HospitalManager.sol 1 ×
contracts > HospitalManager.sol
    }

    // Patients

    struct Patient {
        uint   patientId;
        string patientSsn;
        string patientFName;
        string patientLName;
        string patientGender;
        string patientEmail;
    }

    mapping(uint => Patient) public patients;

    function createPatient( string memory _patientSsn,
                            string memory _patientFName,
                            string memory _patientLName,
                            string memory _patientGender,
                            string memory _patientEmail
        ) public returns (uint) {
        patientCount ++;
        patients[patientCount] = Patient(patientCount,
                                         _patientSsn,
                                         _patientFName,
                                         _patientLName,
                                         _patientGender,
                                         _patientEmail);
        return (patientCount);
    }
```

Fig. 2. Smart contract code for patient registration.

view their health records. The authenticity and access control of the audit trail is managed by passwords that are stored against the entities, such as doctors or staff. These passwords will have to be provided to the smart contract while storing patient visit details such as consultations, surgeries, diagnoses, and prescriptions.

The proposed system tries to create a redundant, tamper-resistant, and verifiable record of timestamped information regarding the patient's medical investigations and health records. It is built on blockchain technology, a distributed, fault-tolerant data store. This information can be used by authorized personnel to locate the corresponding detailed records in the Web UI at each hospital which generates a comprehensive patient treatment file for a thorough investigation. The proposed system uses a consortium blockchain, owned and managed by the government or some central medical authority, to provide secure access to the concerned individuals.

We have tried to stress the ease of creating a simple audit trail based on a chronological record of patient investigations and consultations without modifying the existing systems and practices of the medical institutions and hospitals concerned. This system can pave the way for better integration of medical records and a more structured and standardized approach to shared medical data.

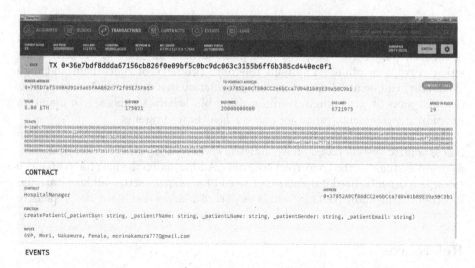

Fig. 3. Smart contract code for patient registration.

5 Challenges and Discussion

We propose the use of blockchain in this system because it is a decentralized database that is immutable. Consider the case of a patient consulting multiple hospitals, each of which maintains its database. No single system has a complete record of all the visits and treatments. Also, consider the case that a particular database is offline or corrupt. The patient or doctor may not know which data has gone missing. In the case of blockchain, as it is a decentralized database with multiple nodes keeping a copy of the data, the entire data will always be available at any given point in time. Also, there is no chance of any corruption or other such failure, which is possible in local databases.

We have proposed the use of Blockchain to eliminate the costs of third-parties and middlemen. However Blockchain technology is inherently complex and expensive; and it would be a challenge to adopt it.

6 Summary and Conclusions

EHR applications include audit trails for auditing and system transparency though each of them can act as a single point of failure. An audit trail has been proposed on Blockchain technology where a patient or a authorized medical personnel has a timeline-based access to all investigations, reports, and procedures conducted during his treatment. The method of facilitating this is through the development of smart contracts, which store the relevant information after providing the necessary checks to ensure the integrity of the data. We have created the Smart Contract for patient registration and validated it in Blockchain. Our proposed approach can meet confidentiality, integrity, and authentication requirements.

A decentralized technology such as Blockchain is used to store EMR and form audit trails. The proposed system can thus solve problems such as fragmentation of medical data across hospitals and possible data loss due to systems being either corrupt or offline. The fault tolerance property of Blockchain and high availability can be utilized to store this information in a distributed manner. This research aims at providing a medium using blockchain technology to maintain an audit management system for current healthcare systems.

For future work, the proposed system can be further enhanced by using an access control system based on asymmetric cryptography with the hospital management systems at the concerned hospitals. Additionally, the hospitals can move towards a decentralized database that would be uniform across all hospitals and incorporate the audit trail blockchain as part of a larger and more comprehensive system.

References

1. Shahnaz, A., Qamar, U., Khalid, A.: Using blockchain for electronic health records. IEEE Access **7**, 147782–147795 (2019)
2. Anshari, M.: Redefining electronic health records (EHR) and electronic medical records (EMR) to promote patient empowerment. IJID (Int. J. Inform. Dev.) **8**(1), 35–39 (2019)
3. Heart, T., Ben-Assuli, O., Shabtai, I.: A review of PHR, EMR and EHR integration: a more personalized healthcare and public health policy. Health Policy Technol. **6**(1), 20–25 (2017)
4. Zhu, H., Hou, M.: Research on an electronic medical record system based on the internet. In: 2018 2nd International Conference on Data Science and Business Analytics (ICDSBA), pp. 537–540. IEEE (2018)
5. Ahmad, A., Saad, M., Njilla, L., Kamhoua, C., Bassiouni, M., Mohaisen, A.: Blocktrail: a scalable multichain solution for blockchain-based audit trails. In: ICC 2019– 2019 IEEE International Conference on Communications (ICC), pp. 1–6. IEEE (2019)
6. Ahmad, A., Saad, M., Bassiouni, M., Mohaisen, A.: Towards blockchain-driven, secure and transparent audit logs. In: Proceedings of the 15th EAI International Conference on Mobile and Ubiquitous Systems: Computing, Networking and Services, pp. 443–448 (2018)
7. Kaushik, K., Kumar, A.: Demystifying quantum blockchain for healthcare. Security and Privacy, p. e284 (2022)
8. Sharma, K., Singh, H., Sharma, D.K., Kumar, A., Nayyar, A., Krishnamurthi, R.: Dynamic models and control techniques for drone delivery of medications and other healthcare items in COVID-19 hotspots. In: Al-Turjman, F., Devi, A., Nayyar, A. (eds.) Emerging Technologies for Battling COVID-19. SSDC, vol. 324, pp. 1–34. Springer, Cham (2021). https://doi.org/10.1007/978-3-030-60039-6_1
9. Kumar, A., Krishnamurthi, R., Nayyar, A., Sharma, K., Grover, V., Hossain, E.: A novel smart healthcare design, simulation, and implementation using healthcare 4.0 processes. IEEE Access **8**, 118433–118471 (2020)
10. Zou, W., et al.: Smart contract development: challenges and opportunities. IEEE Trans. Softw. Eng. **47**(10), 2084–2106 (2019)
11. Truffle Suite: Ganache (2022). https://trufflesuite.com/ganache/
12. Truffle Suite: Ganache overview (2022). https://trufflesuite.com/docs/ganache/

Sentinel: An Enhanced Multimodal Biometric Access Control System

N. Krishna Khanth⬛, Sharad Jain⬛, and Suman Madan⁽⊠⁾⬛

Jagan Institute of Management Studies, Sector-5, Rohini, Delhi, India
ajaykrishna1009@gmail.com, mp.sharadjain24@gmail.com,
madan.suman@gmail.com

Abstract. Every place be it a household or organization, big or small, like banks have something that needs to be secured to ensure efficient operations and management. Security is always a concern as what is being protected is valuable. Security systems based on singular or multiple biometrics such as face, voice, iris, fingerprint and palm along with things being carried in person such as RFID card or security key(s) are used along with or instead of pin, password based existing lock systems is mostly used because of the uniqueness and added layer of security provided by the aforementioned features. But the implementation of these features alone is not sufficient to thwart any malicious actions to gain access to a secure location, due to rise in technology capable of beating/bypassing said security systems. Thus, this paper proposes a robust security system that will be take care of security requirements of any location that might contain something valuable and to be retrofitted with the problems prevailing in the present systems. The proposed system is capable of detecting & recognizing a person's face, their emotion based on facial expression, the liveliness factor of their face to determine physical presence, identifying the speaker along with a word/phrase in their speech and detecting factors in the surrounding environment that may threaten a user. The system is designed in a way such that anyone who wants to enter/access a secure location has to pass through all of these layers like password, facial recognition, facial emotion recognition, facial liveliness recognition, speaker recognition, speaker phrase detection, and environmental threat detection etc. of security working in unison, none of which can be bypassed easily. All the sensors for detecting, identifying and recognizing said biometric features are securely connected to a singular security device to ensure success of this goal.

Keywords: Security · Face · Facial expression and facial liveliness detection · Recognition · Speaker and speaker safety phrase detection · Recognition · Environmental threat detection · Recognition · Smart device

1 Introduction

In general, in today's world everyone is set out to make money, legitimately or by hook or crook. Those who try to make money the wrong way often go down paths like petty theft to grand larceny. To thwart such illicit activities security systems have been put in place, but deviant people come up with ways to circumvent said security systems [1].

S. Sachdeva et al. (Eds.): BDA 2022, LNCS 13830, pp. 95–109, 2023.
https://doi.org/10.1007/978-3-031-28350-5_8

With the help of technology this system contains unauthorized human intervention in an extensive manner. This proposed framework can secure any individuals or organization's valuables placed in bank vault, lockers etc.

When an authorized person tries to access a secure location they have to enter their user ID and password as initial verification start-up which is followed by the person placing themselves in front of the security device which will then use a camera to immediately scan their face, their facial expression, the liveliness status of the face, another camera simultaneously scans the room to detect any elements trying to gain unauthorized access by threat factors, along with a microphone that records the voice of the person and detects/searches for a phrase/word in the voice al happening at the same time. If any of the above-mentioned verifications fail, a security alarm is triggered or access is not granted based on the client's demand.

With the assistance of innovation this framework contains unapproved human mediation in an extensive manner. This framework can be secure any people or association to shield their assets put in bank vault, storage spaces and so on.

2   Motivation

With the existing tech stack, every security centric smart device has facial recognition to prevent unauthorized access. RFID readers and other sensors, such as fingerprint sensors, are also employed in security systems and other applications. Dwi Ana Ratna Wati and Dika Abadianto suggested a smart security system based on facial recognition in 2017 [2]. Based on MyRIO 1900, the security system was created using LabVIEW [2]. When the distance is less than or equal to 240 cm, the system can recognise a person's face, even if they are wearing various items like glasses or a hat. [2]. However, if there is too much change or if the subject is too far away from the camera, it is unable to recognise the face.

The face detection method used in this work is extremely similar to another face recognition security system that Ibrahim Mohammad Sayem and Mohammad Sanaullah Chowdhury built in 2018 utilizing a Raspberry Pi based on Python language and the OpenCV Library [3]. In the event that an unauthorized individual is in front of the camera, the system also emails the authorizer [3]. However, because the system simply relies on facial detection, it is simple to circumvent.

Teddy Mantoro and Suhendi devised an enhanced security system that uses a hybrid approach of the Haar Cascade Classifier to recognize several faces at once [4]. Compared to previous systems, this one is more precise and efficient.

Raj Gusain, Hemant Jain, and Shivendra Pratap created a sophisticated bank security system that uses facial recognition, iris scanning, and palm vein authentication [5]. However, the system is expensive and difficult to maintain. Moreover, Kanza Gulzar, Jun Sang, and Omar Tariq have created and executed a cost-effective Arduino-based solution [6]. However, such gadget takes more time and is unable to identify a human face in a real-time video. Modifying on the work of Kanza Gulzar, Jun Sang and Omar Tariq, Mahadi Hasan Moon, Aurongo Jeb, Mahidul Islam, Debashish Kumar Ghosh, Nafiz Ahmed Chisty, Designed and Implemented a Vault Security System with real time face recognition, fingerprint verification & RFID scanning [16].

Table 1 shows the summary of features included in work done by other authors compared to the proposed system.

Table 1. Comparative study

Paper citation	[2]	[3]	[4]	[5]	[6]	[16]	Proposed
Face recognition	Yes	Yes	Yes	Yes	Yes	Yes	Yes
Facial emotion recognition	No	No	No	No	No	No	Yes
Facial liveliness recognition	No	No	No	No	No	No	Yes
Voice recognition	No	No	No	No	No	No	Yes
Phrase recognition	No	No	No	No	No	No	Yes
Environmental threat detection	No	No	No	No	No	No	Yes
Thumb print scanner	No	No	No	No	Yes	Yes	No
Palm scanner	No	No	No	Yes	Yes	No	No
Iris scanner	No	No	No	Yes	Yes	No	No

3   Proposed System Architecture

The Fig. 1 visualizes the block diagram of the stage 1 of the system. Here, an admin can add new users who are authorized to access the secure location. The admin enters the username of new user and asks the new user to enter a password, select challenge phrases for phrase detection. The system automatically ensures that no field is blank, entered password is strong. If any of the above 2 conditions are not satisfied biometric readings are not started. Although, upon successful completion new user is redirected to the biometrics scanning window. Here the new user can press the button to start recording their voice to train a model for speaker recognition, followed by clicking the button to scan their face to train a model for face recognition. Or the user can start both of them together. If at any point a failure occurs the process is rolled back. After successfully adding a new user to the access list, the personal biometric data stored to train the models are purged and the vault/secure location device is updated to accommodate the newly added user.

The Fig. 2 visualizes the block diagram of the stage 2 of the system. Here, an authorized user will enter their credentials, if credentials are not valid, certain number of retries are allowed, and if all attempts are exhausted security alarms are triggered. If credentials are successfully validated, user is redirected to biometrics scanning window. Here on the click of a button, to ensure real-time security without disrupting the entire system, three separate environments are created.

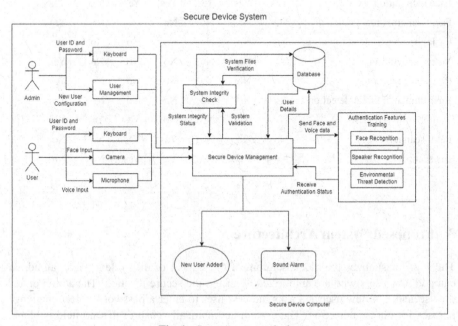

Fig. 1. Stage-1: secure device

First for processing the image to identify the face, facial expression and liveliness status of the person in front of the device, another for recording the voice of the person in front of the device and detecting a phrase/word in it and another for environmental threat detection to ensure the user trying to access the secure location is not being threatened to do so. All of this is happening at the same time. If any of the biometric recognitions fails a security alarm is triggered. If facial emotion reports emotion being fear security alarm is triggered. If facial liveliness fails, i.e., if the person in front of the camera is a recording, a security alarm is triggered. If phrase detector identifies unsafe phrase/word in the speech of the user in front of the device, security alarm is triggered. If environmental threat detector picks up any objects of threat such as gun or knife security alarm is triggered. If all factors return positive results, user is granted access to secure location.

Vault Gate System

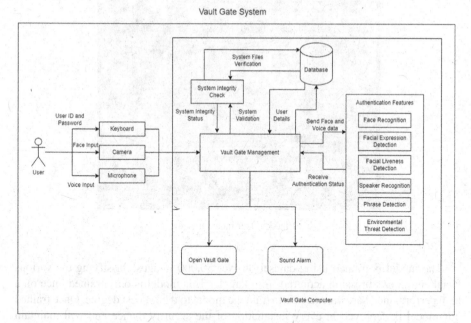

Fig. 2. Stage-2: secure location device

4 Proposed System Design Methodology

4.1 Facial Detection and Recognition Using OpenCV

The user's face is recognized in python language using the OpenCV library. It is a fundamental design blueprint for applications based on computer vision [7]. The OpenCV library contains many algorithms that can and are being used in applications to detect and recognize a face in a frame [7, 18]. Figure 3 shows our face dataset for experimentation. The objective is to detect & identify the face of a human in real time environment. The algorithm used to accomplish this objective is the Haar Cascade Classifier. OpenCV Haar Cascade Classifier operates on two parts: Trainer and Detector [8]. Training of a classifier model is done via the Trainer and the detection of any object, face in this instance is done via the Detector. Images are required to train the classifier. Every image is a grey image of a cropped face of the user from various angles for training the classifier.

4.2 A Facial Emotion Recognition Using CNN Model

The user's facial expression/emotion is recognized in python language using Tensorflow & Keras libraries & packages. It offers a standard framework for applications including & based on machine learning.

The objective is to detect & identify the expression/emotion of the face in real time environment. For this we are using supervised machine learning done in a 4-layer CNN model with 2 layers add on for accurate recognition. The dataset is labelled as the emotion they represent. Figure 4 and 5 shows the happy and fearful datasets sample.

Fig. 3. User face dataset

The model is trained using dataset of over 50,000 images classifying the various facial emotion/expression acquired from kaggle. This model is only trained once on a high performance hardware device to train the model to the highest degree. Once trained this model is deployed in every installation of the security device and will maintain a stable accuracy rate given that the lighting conditions and user facing the camera properly.

Fig. 4. Happy face dataset   **Fig. 5.** Fearful face dataset

4.3 Speaker Recognition Using Fast Fourier Transformation and CNN

The user's speech/voice is recognized in python language using Tensorflow & keras libraries & packages. It offers a standard framework for applications including & based on machine learning.

The objective is to identify & recognize the voice of a speaker from the sounds present in a real time environment. We categorise speakers using speech recordings' frequency domain representation, which we extract using the Fast Fourier Transform. (FFT). [11, 18] To prepare the dataset, speech samples from several speakers are used, and the speaker ID is used as label. To improve the data, background noise is added to these samples before the FFT of these samples is performed. When given a noisy FFT voice sample as input, a 1D convnet is trained to predict the actual speaker [12].

A dataset of 30 s voice/speech recording of a single user is collected and stored temporarily to train the speaker recognition model, which is also unique, different i.e. standalone for every user. This has been done to stabilize the accuracy of the model regardless of the number of users.

4.4 Phrase Detection Using Google Speech to Text API

The user's phrase is detected & identified in python language using Speech Recognition library. It offers a standard framework for applications including & based on machine learning and speech to text based operations.

Google's speech recognition technology in Speech Recognition is used for speech to text conversion. Once text is obtained, it is compared to safe phrase and alert phrase from database.

4.5 Environmental Threat Detection Using OpenCV and YOLO Model

The threat factors are recognized in python language using OpenCV, YOLO libraries & packages. It offers a standard framework for applications including & based on machine learning.

The objective is to identify & detect objects of intimidation used to threaten a user to gain access in a real time environment. The YOLO (You Only Look Once) one-stage object detections in real time environment; YOLOv6 is inspired by YOLOv5 by Ultralytics. The YOLO algorithms are designed adept in object detection. Using machine learning on various images classified as threats, the system learns how to detect threats accurately from frames [10].

5 Experimental Setup and Flow

The secure device allows addition of new users of manager and staff level clearance, it accepts their credentials and reads their biometrics to train & create model files for recognition. It also allows removal of user based on operational hierarchy. Figure 6 shows the simultaneous multi-modal security flow of the proposed system.

The Secure Location Access Device regulates, monitors and authorizes entry of users. It scans their biometrics and check if the user in frame is in duress, present physically. As shown in Fig. 7.

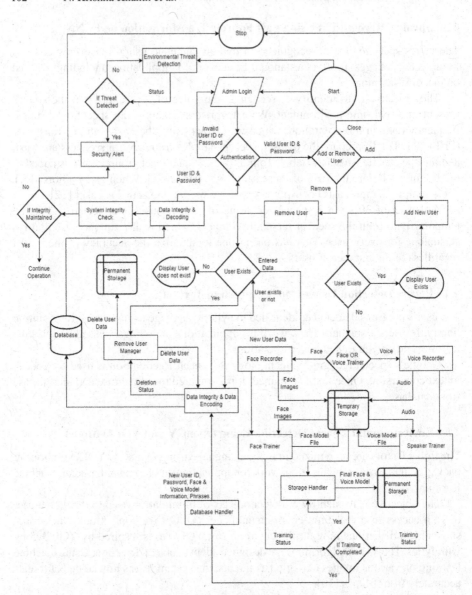

Fig. 6. Secure device flow

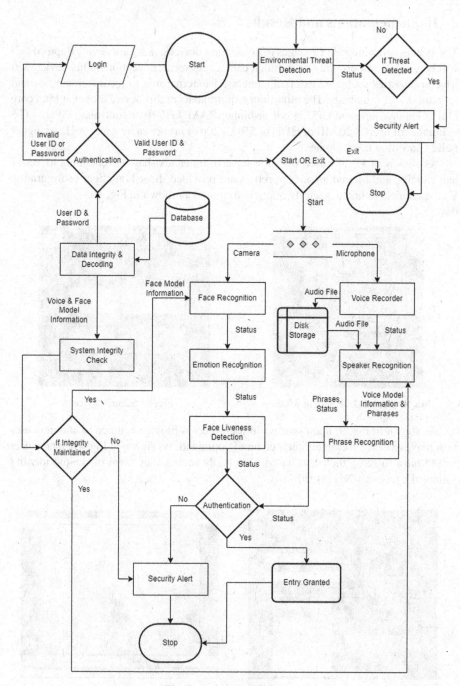

Fig. 7. Secure location flow

6 Implementations and Results

The system is developed in 2 stages of computing devices. A secure device (laptop) that will be used to add or remove users that can access secured locations. This device will have accessories such as 2 external cameras, 1 external mic to record biometrics and monitor the surroundings. The minimum requirements of this device are set at Intel core i3 6th Gen or equivalent CPU, 4 GB or higher RAM, GT710 or equivalent GPU, 4 GB + number of users * 20 MB of HDD or SSD, 720p or higher camera and 90Hz or higher noise cancelling microphone.

As shown in Fig. 8, user will first have to enter text-based credentials to verify authenticity, and limited amount of retries are provided. Based on client configuration if all attempts are failed, security alarm is triggered as shown in Fig. 16.

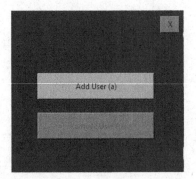

Fig. 8. Secure device login window   **Fig. 9.** Secure device

As shown in Fig. 9, upon successful login user is provided choice to add new user or remove existing user under their chain of command. As shown in Fig. 10, a new user would have to setup their text-based credentials such as any form of unique identity followed by a password [9, 15].

Fig. 10. Credentials setup window   **Fig. 11.** Face & voice training window

Strong password policies are enforced and will prevent further operation if a weak password is entered. No filed can be left blank, criticalities like words in phrase selection are pre-set and varying for every user to avoid conflict and no word in the safe phrase set is present in the unsafe phrase set and vice versa. Than as shown in Fig. 11 the user can then click the button to scan their face to train a model for face recognition, or click the button to record a speech sample to train the speaker recognition model as shown in Fig. 12, or the user can start both of the scanning & recording processes together. If at any point a failure occurs the process is rolled back.

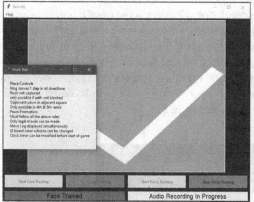

Fig. 12. Face & voice training window   **Fig. 13.** Secure location access device

Upon successful completion of training the new user setup is finished and all their provided data is stored, but the personal biometric data stored to train the models are purged. If there is failure at any point all the progress is rolled back. A secure location access device (laptop) that will be used by authorized users to gain access to secure location. This device will have accessories such as two external cameras, one external mic to record biometrics and monitor the surroundings. The minimum requirements of this device are set at Intel core i3 6th Gen or equivalent CPU, 4GB or higher RAM, GT710 or equivalent GPU, 4 GB + number of users * 20 MB of HDD or SSD, 720p or higher camera and 90 Hz or higher noise cancelling microphone.

Fig. 14. Biometric verification window   **Fig. 15.** Recording, validation of biometrics

As shown in Fig. 13, user will first have to enter text-based credentials that were setup during the registration process to verify authenticity, and limited amount of retries are provided for failures. Based on client configuration if all attempts are failed, security alarm is triggered as shown in Fig. 16. As shown in Fig. 14, upon successful login user has to provide biometric data to gain access by pressing a button.

As show in Fig. 15, when the user starts the biometric verification process by clicking the button in Fig. 14, they have to show their face and speak some sentences which contain the phrase selected by them during the registration process and if the face, voice are recognized, phrase/word in the users speech doesn't contain the unsafe phrase selected by them and it is also verified that the user is not in duress i.e. being threatened in any way by recognizing their facial emotion/expression along with scanning the surroundings to identify any objects of threat such as gun or knife, and that user is actually physically present and the face and voice being processed is not a recording or digital spoof in any way, then they are allowed access. If any of the verification process fails, based on client configuration, security alarm is triggered as shown in Fig. 16.

Fig. 16. Security measures triggered

6.1 Registration Time

The time taken for a user to setup their credentials, system to scan their biometrics and train models. It takes up to 4 min to successfully complete the process.

Total Time Taken = Time to setup credentials (User Dependent)
+ Time to scan face (Constant)
+ Time to train face model (System Dependent)
+ Time to record voice (Constant)
+ Time to train voice model (System Dependent).

6.2 Entry Time

The time taken for a user to enter their credentials, provide their biometrics for the system to recognize and gain entry. It takes up to 15 s to successfully validate a user's authority to access secure location.

$$\text{Total Time Taken} = \text{Time to enter credentials (User Dependent)}$$
$$+ \text{Time to scan face (Constant)}$$
$$+ \text{Time to recognize face (System Dependent)}$$
$$+ \text{Time to record voice (Constant)}$$
$$+ \text{Time to recognize voice (System Dependent)}$$

6.3 Model Accuracy

The various machine and deep learning models used for validation of a user's biometric features provide a satisfactory level of accuracy to assure quality of security. The face recognition based on OpenCV Harcascade model provides an accuracy range as shown in Table 2 [S1]. This slight variation in accuracy is observed due to lighting conditions and how the user is facing the camera. The facial emotion/expression recognition based on CNN model provides an accuracy range as shown in Table 2 [S2]. This variation in accuracy is observed due to the quality of camera, dataset used in training of the model, and inability of a user to maintain a constant facial expression.

The facial liveliness based on MobileNet & CNN models provides an accuracy range as shown in Table 2 [S3]. This variation in accuracy is observed due to the quality of camera, dataset used in training of the model, lighting conditions and how the user is facing the camera. The speaker/voice recognition based on CNN model provides an accuracy range as shown in Table 2 [S4]. This variation in accuracy is observed due to the quality of microphone used, and intensity of background noise.

Table 2. Accuracy of models

SNo	Feature	Accuracy (%)
[S1]	Face recognition	85–92
[S2]	Facial emotion/Expression recognition	80–85
[S3]	Facial liveliness detection	70–75
[S4]	Voice/Speaker recognition	85–90

6.4 Privacy Aspect

Personal biometric data taken for training machine learning models are not stored permanently, keeping in mind a user's privacy expectations. All other data & models are encrypted using various cryptographic techniques & k-anonymity for tamper proofing [13, 14, 17].

7 Conclusion

The objective is to provide a satisfactory experience and peace of mind regarding safe-keeping. The system implements multi-factor authentication to obtain results. Although deployment cost is high as high end hardware is used for maximum security. It improves security by limiting any kind of possible security breaches and threats. The entire project has been developed & designed in python language keeping platform independence in mind. More biometric features such as palm print, iris/retina scanners can be added to improve security layers.

References

1. Murugan, K.H.S., Jacintha, V., Shifani, S.A.: Security system using raspberry Pi. In: 2017 Third International Conference on Science Technology Engineering & Management (ICONSTEM), Chennai, pp. 863–864 (2017)
2. Wati, D.A.R., Abadianto, D.: Design of face detection and recognition system for smart home security application. In: 2017 2nd International conferences on Information Technology, Information Systems and Electrical Engineering (ICITISEE), Yogyakarta, pp. 342–347 (2017)
3. Sayem, I.M., Chowdhury, M.S.: Integrating face recognition security system with the internet of things. In: 2018 International Conference on Machine Learning and Data Engineering (iCMLDE), Sydney, Australia, pp. 14–18 (2018)
4. Mantoro, T., Ayu, M.A., Suhendi: Multi-faces recognition process using haar cascades and eigenface methods. In: 2018 6th International Conference on Multimedia Computing and Systems (ICMCS), Rabat, pp. 1–5 (2018)
5. Gusain, R., Jain, H., Pratap, S.: Enhancing bank security system using face recognition, iris scanner and palm vein technology. In: 2018 3rd International Conference on Internet of Things: Smart Innovation and Usages (IoT-SIU), Bhimtal, pp. 1–5 (2018)
6. Gulzar, K., Sang, J., Tariq, O.: A cost effective method for automobile security based on detection and recognition of human face. In: 2017 2nd International Conference on Image, Vision and Computing (ICIVC), Chengdu, pp. 259–263 (2017)
7. Patoliya, J.J., Desai, M.M.: Face detection based ATM security system using embedded Linux platform. In: 2017 2nd International Conference for Convergence in Technology (I2CT), Mumbai, pp. 74–78 (2017)
8. Kim, M., Lee, D.G., Kim, K.-Y.: System architecture for real-time face detection on analog video camera. Int. J. Distrib. Sens. Netw. **11**, 1–11 (2015)
9. Chen, Y., Chen, Z., Xu, L.: RFID system security using identity-based cryptography. In: 2010 Third International Symposium on Intelligent Information Technology and Security Informatics, Jinggangshan, pp. 460–464 (2010)
10. Shaikh, S., Raskar, R., Pande, L., Khan, Z., Guja, S.P.: Threat detection in hostile environment with deep learning based on drone's vision. In: 2020 International Research Journal of Engineering and Technology (IRJET) (2020)
11. Oran Brigham, E.: The Fast Fourier Transform and Its Applications. Prentice-Hall Inc., Hoboken (1988)
12. Li, R., Jiang, J.Y., Liu, J., Hsieh, C.C., Wang, W.: Automatic speaker recognition with limited data. In: Proceedings of the 13th International Conference on Web Search and Data Mining (WSDM 2020), pp. 340–348. Association for Computing Machinery, New York 2020. https://doi.org/10.1145/3336191.3371802

13. Madan, S., Bhardwaj, K., Gupta, S.: Critical analysis of big data privacy preservation techniques and challenges. In: Khanna, A., Gupta, D., Bhattacharyya, S., Hassanien, A.E., Anand, S., Jaiswal, A. (eds.) International conference on innovative computing and communications. AISC, vol. 1394, pp. 267–278. Springer, Singapore (2022). https://doi.org/10.1007/978-981-16-3071-2_23

14. Madan, S., Goswami, P.: Hybrid privacy preservation model for big data publishing on cloud. Int. J. Adv. Intell. Paradigms **20**(3–4), 343–355 (2021). https://doi.org/10.1504/IJAIP.2021.119022

15. Boopalan, S., Ramkumar, K., Ananthi, N., Goswami, P., Madan, S.: Implementing ciphertext policy encryption in cloud platform for patients' health information based on the attributes. In: Singh, V., Asari, V.K., Kumar, S., Patel, R.B. (eds.) Computational Methods and Data Engineering. AISC, vol. 1227, pp. 547–560. Springer, Singapore (2021). https://doi.org/10.1007/978-981-15-6876-3_44

16. Moon, M.M.H., Ghosh, D.K., Chisty, N.A., Jeb, M.A., Islam, M.M.: Design and implementation of a vault security system. In: 1st International Conference on Advances in Science, Engineering and Robotics Technology 2019 (ICASERT 2019) (2019)

17. Puri, V., Kaur, P., Sachdeva, S.: Effective removal of privacy breaches in disassociated transactional datasets. Arab. J. Sci. Eng. **45**, 3257–3272 (2020)

18. Madan, J., Madan, S.: Intelligent and personalized factoid question and answer system. In: 2022 10th International Conference on Reliability, Infocom Technologies and Optimization (Trends and Future Directions) (ICRITO), pp. 1–7 (2022). https://doi.org/10.1109/ICRITO56286.2022.9964818

Big Data Analytics in Healthcare Support Systems

A Programmatic Solution to Stop Heartbleed Bug Attack

Urvashi Chugh[1] (ID), Amit Chugh[2] (ID), Prabhakar Agarwal[3(✉)] (ID),
and S. Pratap Singh[4] (ID)

[1] IT Department, KIET Group of Institutions, Delhi NCR, India
[2] CSE Department, IMS Engineering College, Ghaziabad, India
[3] CSE Department, National Institute of Technology, Delhi, India
prabhakar@nitdelhi.ac.in
[4] GCET, Greater Noida, India

Abstract. A flaw was found in the Open SSL cryptography library in April 2014, known as the Heartbleed vulnerability that was implemented in the Transport Layer Security and Secure Socket Layer Protocols. This bug allowed the attacker to steal sensitive data from the victim's memory servers. This vulnerability was present on many web servers and major sites, including Yahoo. Many servers could have a significant loss due to this. This research paper has discussed the Heartbleed vulnerability and proposed one solution to fix this for developer security. The Objective is to find a programmatic solution for heartbleed vulnerability to prevent the victim from losses. This proposed work has a major impact on authenticity and security while using open-source projects. This research paper will present a coding way of checking payload length before transferring the data to fix this bug.

Keywords: Heartbleed · Open SSL · Transport layer attack

1 Introduction

The heartbleed formal name CVE-2014-0160 given by Codenomicon, is a vulnerability that gives a chance to the attacker to connect to the client's server or memory without any authentication [1]. This extension helps maintain user Internet connections without the need for continuous data transfers. Heartbleed was caused by poorly written code. Security researchers identified that heartbleed could enable an intruder to attack by exploiting passwords, user names, encrypted keys, etc. Since 1998 open SSL is widely used by many active websites on the internet, many security researchers [2] have called it one of the most dangerous security bugs in the history of the internet. It was first detected by Neel Mehta of the google security team, and later on 7th April 2014, it was formally included in CVE. An example of an unsafe Open SSL Heart Bleed can be taken as a Transport Layer Security Server or a Storefront. This leads to incorrect input validation in the implementation of the transport layer security heart rate extension. Hence, the name harm comes from heart pain. Vulnerability is assessed as a reading buffer; There are often cases where additional knowledge is allowed. This paper had objectives of

S. Sachdeva et al. (Eds.): BDA 2022, LNCS 13830, pp. 113–121, 2023.
https://doi.org/10.1007/978-3-031-28350-5_9

introducing heartbleed vulnerability it causes and introducing a programmatic solution for the prevention of this attack. Heart failure is reported as a common weakness and exposure to CVE-2014-0160. The Open SSL hard and fast version was released in 2014 and on the same day, it was exposed to blood, meaning that the data was made vulnerable and security information could leak from the client's memory, and many of the Transport Layer Security enabled websites are still likely to bleed. The vulnerability CVE-2014-0160 leaves behind no trace of attacks in the server log files, so the clients have no information about the attack until and unless the information leaked attacks them negatively. While using open-source codes [3], it is essential for the user for been guaranteed user security and data confidentiality. Heartbleed bug is a bug on the Open SSL technology, which implements both SSL (Secure Socket Layer) and TLS (Transport Layer Security) [4]. The bug is located in the heartbeat extension of the Open SSL. RFC6520 defines SSL Heartbeats extensions that are used to keep a connection alive without the need to constantly renegotiate the SSL session. When it is exploited, it causes a memory leak in the contents from the client to the server and vice versa. That is why it is called the Heartbleed bug [5]. This work has resolved the vulnerability by verifying the asked payload length with the length of data being transferred.

It has been many years since the Heartbleed flaw was discovered but still today it can be found on certain open servers and managing security in such a vast infrastructure is uncannily difficult and is like managing pests on farms. Handling them with proper rectification can help in reducing the issue area whereas neglecting it might turn out to do a lot of harm to the system and might take a lot of work to rectify [6]. This Heartbleed vulnerability was a coding error that opened several doors for cyber criminals that helped them steal private information from the victim's server and memory and this leaked the victim's personal details such as usernames, passwords, token keys, etc. The attackers could misuse the bug and access sensitive data and information while targeting the server's memory.Heartbeat extension was launched in 2012 and the heartbleed bug was discovered in 2014 and security experts [7] imply that a lot of private information and sensitive data might have been leaked in this time period. Heartbeat extension was added to check connection health. Attackers misused this heartbeat extension and leaked crucial data [8].

Open-source projects have little financial support for purpose of review or upgrades. It left abundant of the world's infrastructure susceptible to cyber criminals. Keeping systems secure needed system directors to not solely update software packages, but however get new master keys to reestablish their company's electronic identity in several cases they conjointly had to raise their users to vary passwords. It's seemingly that the worldwide value of handling Heartbleed has already run into the many countless greenbacks. A second spherical of issues within the month of 2014, once more requiring respectable remedial action by vendors and system directors.Most of the vulnerable systems are now secure, however, some like the older smartphones having a vulnerable code are not secured. To secure these devices, the user had to update the phone to the latest versions or the user could either patch the provided version to repair the vulnerability [9]. Thus, vulnerable phones ought to be updated to guard them. Google created updates out there to the makers of smartphones shortly once discovering the matter however makers then had to use Google's fixes to the particular microcode for every of their affected models, and take a look at the fastened version. Even then, updates for several phones weren't created out there to shoppers, as phones area units usually oversubscribed with bespoke Vodafone offer custom microcode in phones oversubscribed from their shops. Every carrier would then have had to package and take a look at the update for the bespoken version for every vulnerable phone model. The internet has become a worldwide necessity and with the increasing internet world, the number of cyber crimes is increasing very fast. To overcome these issues the world requires more secure protocols to protect user data and prevent criminal activities [10].

Security over the Internet can be applied layer and layer. Each layer is having applicable protocols. Network layer security is one of them. Security over the network layer of TCP/IP model can be improved by using the protocol Secure Socket Layer (SSL) or its follow-on protocol Transport Layer Security (TLS). The Secure Socket Layer or SSL is a technology for safeguarding sensitive data and keeping a secure internet connection between 2 systems to protect the systems from an external attacker and keep the data encrypted. This protocol makes it difficult to decipher any data that is being transmitted between any devices. SSL protocol applies security at the transport layer.SSL protocol is a combination of further four protocols SSL handshake, SSL change Cipher, SSL alert, and SSL record protocol. SSL handshake protocol is the most important protocol that helps to establish a secure connection between client and server. It helps in negotiation and exchanging capabilities and certificates. Handshake protocol initiates with the exchange of a hello request message. Upon receiving the hello request client and server send a hello message including parameters like version, keys, compression methods, and session id, etc. After this certificates are exchanged. Through certificates, client and server can verify the identity of each other. Handshake protocol provides the agreed keys to record protocol for encryption. SSL record protocol ensures confidentiality by encryption and integrity by MAC (message authentication code). It first fragments the received data from the application layer. Fragmentation is done into fixed-size blocks of size 16,384 bytes. In the next step, SSL record protocol compress these block to reduce the size to 1024 bytes. After compression adding MAC is the next step. MAC is inserted code that is strictly dependent on the original message to ensure integrity. In case due to attack, any byte is changed in the original data then MAC will not match. Further record protocol will encrypt and add an SSL record header. SSL change cipher protocol is to resume the session state in case left earlier. SSL alert protocol performs the task of sending alert and error messages. Transport Layer security or TLS is just an advanced version of SSL. This widely used protocol is designed to facilitate privacy and data security for communications over the internet. The communications between web applications and servers can be encrypted using this protocol.

These two protocols are commonly referred to together as SSL/TLS [11]. The Heartbleed vulnerability of OpenSSL is widely used SSL and TLS protocol. This leads to incorrect input validation in the implementation of the transport layer security heart rate extension. The OpenSSL software library contains open-source implementations of Secure Socket Layer (SSL) and Transport Layer Security (TLS). The core library is written in C programming language which provides basic cryptographic and utility functions. OpenSSL versions 1.0.1 through 1.0.1f had a severe memory handling bug in their implementation of the transport layer security Heartbeat Extension that could be used to reveal up to 64 KB of the application's memory with every heartbeat. And with this memory revealing, the attacker could access sensitive information like the server's private key, and the use of this encrypted key might help the attackers to decode the communications, and didn't ensure perfect security. Normal usage is shown in Fig. 1 and Malicious usage is shown in Fig. 2.

Fig. 1. Normal use

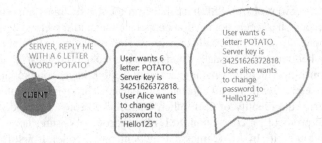

Fig. 2. Malicious use

2 Literature Review

The author Wheeler, David's research makes the world aware of the vulnerability of heartbleed. This paper introduced and discussed how exchanged SSL messages can have improper validation and system can be forged to get over privilege access. Heartbleed is the result of buffer over read problem. Buffer over read caused due to improper input validation. Many servers failed in Input validation and the manual test failed to find the reason for this heartbleed attack. This paper has discussed many approaches to recognize and counter this Heartbleed vulnerability. Negative testing is one method to check input validation. In negative testing, wrong inputs are given intentionally to check whether they are rejected or not. Fuzzing is one more way to detect heartbleed. In the fuzzing method, it is to detect whether the illegal memory access is happening. Tool tries to read out of bound addresses. Fuzzing can also be done on output. One more tool source code weakness analyzer can be used to detect the weakness of software. Some tools can also check full branch implementation. The author wheeler eventually develop software that helps in identifying this vulnerability. The architecture model talk in this model was very unique and commendable [12]. The authors Wang, Jun, et al. have focused on working on the idea of the vulnerability of buffer overflow they tend to make a model with will detect whether the system's confidential data is safe or not [13].

The authors ghafoor et al. in paper [14] talked about the evolution of embedded systems to the Internet-of-Things where every device will be connected to the internet. This scenario required big security more specifically information security. This paper has focused on the security of embedded devices again heartbleed. Once it's found that open SSL is prone to heartbleed, then next target can be embedded systems because they were also relying on the security of SSL. Embedded system was growing with the support of advanced microprocessors and real-time operating systems. Later on, development of ARM processor-compatible operating systems needs more security from where the use of SSL started. In that era, Open SSL was being used by more than two third of web servers including famous brands like Google.

The authors M.Carvalho, J.DeMott, R.Ford, D.Wheeler look at this vulnerability in OpenSSL and outline how it was fixed. They also address why the Heartbleed vulnerability was missed for so long [15].

3 Proposed System

It's a known fact that SSL and TLS protocols provide security and privacy for communication over the internet for applications like IMs, e-mails, and VPNs. As observed earlier, because of the Heartbleed vulnerability any attacker can scan the memory of the server system protected by vulnerable versions of SSL and TLS that might compromise the key, token names, encrypted data, etc. This vulnerability is not in the protocols of SSL and TLS, but it is a programming mistake in the well-liked OpenSSL library. The proposed system makes it capable of simulating planet attacks on the Open SSL version one.0/1.0.1. As security ought to be a priority for everybody, it tends to aim at building a userfriendly tool that might be utilized by professionals and new folks alike, to uncover the vulnerabilities in their sites.

3.1 Types of Messages

Heartbleed uses two types of messages one is *HeartbeatRequest* and the second is *HeartbeatResponse*. *HeartbeatRequest* contains the request message that the user might request from the server. Heartbeat Response contains the response message sent back by the server. *HeartbeatRequest* message can arrive at any time during the lifetime of a connection and not more than one HeartbeatRequest message can be there at the time of flight i.e. at the time when one HeartbeatRequest message is received or till the time the transmission of one HeartbeatRequest message expires. The following is the algorithm for HeartbeatRequest and HeartbeatResponse messages:

```
struct
{ HeartbeatMessage type;
payload_length;
payload[HeartbeatMessage.payload];
padding[padding_length];
}HeartbeatMessage;
```

Here in the above algorithm, type contains the type of message being transmitted i.e. a request message or a response message.payload_length contains the length of the message that is to be sent. Payload contains the actual message. Padding contains any random message.

3.2 Rectified Code

Here is a look at the coding mistake that caused the Heartbleed Attack:

memcpy(bp, pl, payload).

Here memcpy() is the command that helps in copying the data from the server, bp is the destination place where the command is being copied to, pl is the source place from where the command is being copied and the payload is the command that is being transmitted.

Being a part of the OpenSSl library, many had access to this code and many might have even seen this code, but no one noticed the error in coding. The error here is that there is no code to check whether the length of data in the payload is equal to the length of data being carried from the pl (source of transmission of data).

As soon as the vulnerability was detected, patches were rolled out for the upgradation of the OpenSSL version. Companies often after using open source tools and create software, the most important thing that can be done is that first check if there is any error in the open source code. The developer should try different tools and techniques as different terminologies work differently and that might be helpful in creating a secure environment. The Heartbleed code was later rectified where it now checks if the length of the command being transmitted is equal to the length of the command sent from the source server.

```
{
    payload_length // Read the payload length first //
if (1 + 2 + 16 > s->s3->relent) // check if message is greater than 0 KB //
return 0;
hbtype = *p++;
n2s(p, payload);
If (1 + 2 + payload + 16 > s->s3->rrec.length) // check if the length of the message is
equal to the length of payload //
return 0;
pl = p;
}
```

3.3 Flow Chart

The first part of the code detects if the message is greater than 0 KB or else it might cause problems and the second part checks if the length of the message requested is equal to the length of the message being communicated. If you discover that the server has been left uncontrolled to vulnerabilities, change the passwords from your system, update your OpenSSL, and also change the SSL certifications used by your server to leave no trace. The flow chart is shown in Fig. 3.

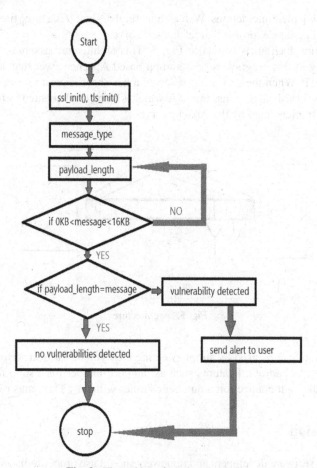

Fig. 3. Flow chart of proposed solution

4 Architecture and Implementation

The interactions between the system and the surroundings help in designing the system architecture shown in Fig. 4. Initially determine the parts that conjure the main system

Fig. 4. Interaction between client and site

and then develop their interactions. Web architectural design: It is a three tire architecture consisting of client side, protocol side, and site server.

Architecture diagram is shown in Fig. 5. This architecture assumes that the web server running in this program is on a xampp based Apache server that supports both https and HTTP. When the server has received a heartbeat request message, the server then reverts the heartbeat response message which is being implemented by the OpenSSL library being implemented by the Apache server.

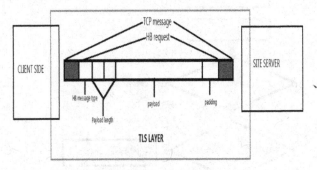

Fig. 5. Architecture

The program can be used by simply running the Python platform script concerning the target server's address. Features such as the port number for testing the TCP port and running the built connection n number of times without TTL counts running out.

5 Conclusion

Conclusively software developers and reviewers should use proper tools and terminologies to test the software for user safety purposes. May technological companies have pledged to take initiative in strengthening the core infrastructure. In this fast-moving technological world, it is time for some introspection for tackling the big stain of heartbleed in the history of compromising security over the internet. The trust in this security vulnerability has broken and only time will help in healing the damage caused. Now it is time to own responsibility and take a step towards securing the systems to protect the information from attackers and cybercriminals. Careful monitoring of critical open source projects like OpenSSL (which depends largely on donations) is important, and code reviews should be regular events, especially before major changes are merged into the main production branch [16]. Security analysts should educate about the vulnerability of all users and victims. When enterprises and other individuals work in a cohesive manner, a stronger open-source security system can be created that will be helpful for all users in the future. If this private information is not secured properly, users of these Open source sites may become the victim of identity fraud and identity theft.

References

1. Heartbleed Keeps Flowing - Open Source Security Melissa Iori (miori01)
2. Sachdeva, S., Mchome, S., Bhalla, S.: Web services security issues in healthcare applications. In: 2010 IEEE/ACIS 9th International Conference on Computer and Information Science, Yamagata, Japan, pp. 91–96. IEEE (2010). https://doi.org/10.1109/ICIS.2010.134
3. Sachdeva, S., Batra, S., Bhalla, S.: Evolving large scale healthcare applications using open standards. Health Policy Technol. **6**, 410–425 (2017). https://doi.org/10.1016/j.hlpt.2017.10.001
4. Bug, T.H.: The heartbleed bug (2021)
5. Yapri, J., Hananto, R.: Leak in OpenSSL. Department of Information Technology, Swiss German University, Tangerang 15143, Indonesia
6. Durumeric, Z., et al.: The matter of heartbleed. In: Proceedings of the 2014 Conference on Internet Measurement Conference, pp. 475–488 (2014)
7. Jain, L., Katarya, R., Sachdeva, S.: Opinion leader detection using whale optimization algorithm in online social network. Expert Syst. Appl. **142**, 113016 (2020). https://doi.org/10.1016/j.eswa.2019.113016
8. Banks, J.: The Heartbleed bug: Insecurity repackaged, rebranded and resold. Crime Media Cult. **11**(3), 259–279 (2015)
9. Kyatam, S., Alhayajneh, A., Hayajneh, T.: Heartbleed attacks implementation and vulnerability. In: 2017 IEEE Long Island Systems, Applications and Technology Conference (LISAT), pp. 1–6. IEEE (2017)
10. Carvalho, M., DeMott, J., Ford, R., Wheeler, D.A.: Heartbleed 101. IEEE Secur. Priv. **12**(4), 63–67 (2014)
11. A technical view of theOpenSSL 'Heartbleed'vulnerability A look at the memory leak in the OpenSSL Heartbeat implementation Bipin Chandra
12. Wheeler, D.A.: Preventing heartbleed. Computer **47**(8), 80–83 (2014). https://doi.org/10.1109/MC.2014.217
13. Wang, J., et al.: Risk assessment of buffer "Heartbleed" overtead vulnerabilities. In: 2015 45th Annual IEEE IFIP International Conference on Dependable Systems and Networks. IEEE (2015)
14. Ghafoor, I., Jattala, I., Durrani, S., Tahir, C.M.: Analysis of OpenSSL heartbleed vulnerability for embedded systems. In: 17th IEEE International MultiTopic Conference 2014, pp. 314–319 (2014)
15. Carvalho, M., DeMott, J., Ford, R., Wheeler, D.A.: Heartbleed 101. IEEE Security Privacy **12**(4), 63–67 (2014)
16. Wheeler, D.A.: How to Prevent the next Heartbleed, 2020-07-18 (originally 2014-04-29)

A Short Review on Cataract Detection and Classification Approaches Using Distinct Ophthalmic Imaging Modalities

Aakash Garg⬤, Jay Kant Pratap Singh Yadav(✉)⬤, and Sunita Yadav⬤

Ajay Kumar Garg Engineering College, Ghaziabad, India
er.jaykant@gmail.com

Abstract. A cataract is one of the leading causes of visual impairment worldwide compared with other major age-related eye diseases, including blindness, such as diabetic retinopathy, age-related macular degeneration, trachoma, and glaucoma. Cloudiness in the lens of an eye leads to an increasingly blurred vision where genetics and aging are the leading cause of cataracts. In recent years, various researchers have shown an interest in developing state-of-the-art machine learning and deep learning techniques-based methods that work on distinct ophthalmic imaging modalities aiming to detect and prevent cataracts in the early stage. This survey highlights the advances in machine learning and deep learning state-of-the-art algorithms and techniques applied to cataract detection and classification using slit lamps, fundus retinal images, and digital camera images. In addition, this survey also provides insights into previous works along with the merits and demerits.

Keywords: Cataract · Visual impairment · Ophthalmic imaging modalities · Machine learning · Deep learning

1 Introduction

The capability of having the sense of sight, eyesight is the utmost part of our senses and is crucial at every turn or point of our life. Without vision, many of us would usually struggle to read, to participate in any work or activity. However, unfortunately, there is a huge community of citizens all over the world hurting from visual impairment including blindness. Visual destruction occurs when an eye becomes unable to see objects as clearly as usual. This condition affects visual perception and one or more of its eyesight functions. The report of the World Health Organization (WHO) says that there are many other age-related eye diseases that aim for visual destruction including blindness, such as diabetic retinopathy, age-related macular degeneration, trachoma, and glaucoma in which cataract is one that is responsible for leading causes of visual destruction and worsens daily. At this time, not less than 2.2 billion people be in pain from visual destruction or blindness and the numbers are rising, of whom at least 1 billion have been avoided and still have a vision (or visual) impairment due to the reasons like poverty,

© The Author(s), under exclusive license to Springer Nature Switzerland AG 2023
S. Sachdeva et al. (Eds.): BDA 2022, LNCS 13830, pp. 122–134, 2023.
https://doi.org/10.1007/978-3-031-28350-5_10

undeveloped countries, and areas with insufficient medical facilities and lack of trained specialists that could have been taken care of and has yet to be addressed [1].

Figures 1(a), and 1(b), show statistics based on the aims for visual destruction worldwide and in India. These two figures depict that the primary reason for an eye disorder is cataracts [2]. Cataracts are associated with many factors, like smoking, aging, high blood pressure, obesity, past eye disease or inflammation, family history of cataracts, and diabetes. Depending on the occurrence of the cataract in the lens of an eye, cataracts may be classified straight into Nuclear Cataracts (NC), Cortical Cataracts (CC), Post Subcapsular Cataracts (PSC), and Congenital Cataracts (CC) shown in Fig. 2 [3]. Accounting for the severity of the cataract, it can be classified into 4-class classification: normal, mild, moderate, and severe grades are detailed for optical eye images. The cataract can be prevented, and major benefits can be attained if it can be discovered and recognized in an early stage before it may result in lifelong vision loss. Hence, this introduces some challenges for scholars and researchers to reinforce a system by applying the set of most suitable algorithms and techniques that accurately detect and grade cataracts.

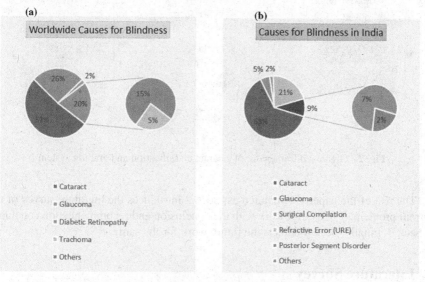

Fig. 1. (a) Details of blindness worldwide [4]. (b) Details of blindness in India [5]

In the past years, ophthalmologists and trained specialists have worked with the manual process of using distinct ophthalmic imaging modalities, mainly with the slit lamp imaging modality to encounter cataracts lying on their medical expertise. This companion approach used to detect cataracts is a costly, time-consuming process and requires a face-to-face consultation which leads to a great challenge for ophthalmologists. In the last few years, researchers show an interest and made efforts to emerging the computer-assisted-detection methodologies [6] and came up with a lot of modern machine learning algorithms like SVM which is a strong classification method, and deep learning algorithms like CNN, Faster-RCNN, Multilayer Perceptron Neural Networks (MLP), Transfer Learning (TL) for automatic cataract detection and classification that give the ophthalmologists a great relief in identification and prevention of the cataract. Figure 2 depicts the methodologies with distinct ophthalmic imaging techniques mostly used for the classification and grading of cataracts.

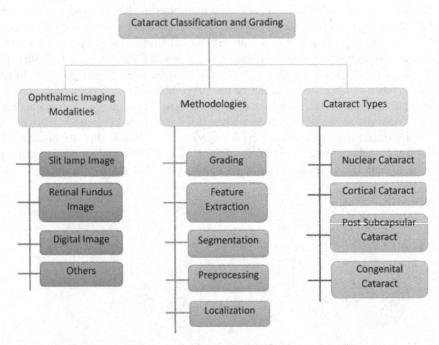

Fig. 2. Organized framework of cataract classification and grading system

The rest of the paper is structured as: Sect. 2 introduces the literature survey of the current problem. The overall work with the conclusion and recommendation concludes in Sect. 3. Finally, Sect. 4 depicts the future work for the same.

2   Literature Survey

After looking at the present and ongoing journals and editions to gather a preferential understanding of the issue and conversing feasible solutions to cataract detection and classification system. As a result, we got conception that several studies have focused on the machine learning and deep learning approaches using slit-lamp imaging modality for cataract detection and classification which is a cost-effective and time-consuming process, and several studies focus on the use of fundus retinal images and digital images for the same. As per several results, nowadays, fundus imaging is broadly used to discover the classification and detection of cataracts that involves four major steps: preprocessing, features extraction, feature selection, and classifier displayed in Fig. 3. This review focuses on showing the various modernized algorithms of machine learning and deep learning with distinct ophthalmic imaging techniques for cataract detection and classification. I hope this review can add great knowledge to the work by providing a valuable summary of the current works and giving a new research direction for the detection and classification of cataracts in the future.

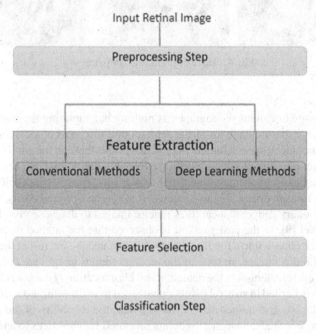

Fig. 3. Major steps used in automatic cataract detection and classification system

2.1 Analysis of Ophthalmic Imaging Modalities for Cataract Detection and Classification

For a better understanding, this section defines three distinct ophthalmic images used in most of the works of literature for cataract classification: slit lamp image, fundus retinal image, and digital camera image. The current section introduces each image type step by step along with the merits and demerits.

Slit Lamp Image
A slit lamp ophthalmoscope is nothing but a diagnosis tool that consists of a high-intensity light source used during the examination of an eye. This slit lamp ophthalmoscope gives an ophthalmologist a detailed view of the different structures of an eye [7]. This process is also used to diagnose cataracts that require a face-to-face consultation. This gadget is quite valuable and demands a great specialization to adopt it which shows the limitation of this imaging modality. Figure 4 represents the different slit-lamp images.

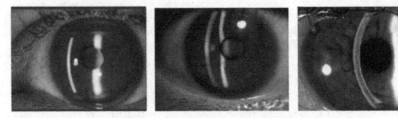

Fig. 4. Three different slit-lamp images

Fundus Image

The fundus image or fundus photography is nothing but capturing the inside, the back surface of an eye that is made up of the retina, optic disc, and blood vessels. Fundus photography includes a special fundus camera that points through the pupil to the back of the eye and takes photographs that are operated by ophthalmologists or trained specialists [8]. Today, Fundus Retinal Images are used to diagnose eye-related problems with the availability of non-mydriatic fundus cameras, that can be operated easily, and are said to be a revolutionary gadget that captures precise images to diagnose visual destruction at the prior level [9]. In the past years, it is observed that the method of conversion of mobiles or smartphones into a fundus camera [10] elevated the interest of the researchers to work with fundus images. In spite of the fact that fundus images are quite functional but having several challenges in the detection and classification of cataracts as the data of fundus images is limited in size, relatively unbalanced to work with, and needs the advice of ophthalmologists and trained specialists to assess the reliability of the diagnosis of fundus images. In a study, a batch of scholars proposed two practices to analyze fundus images (Fig. 5).

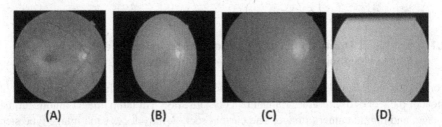

(A)   **(B)**   **(C)**   **(D)**

Fig. 5. Pictures of fundus image with different levels of cataracts lying on their severity (A) Normal (B) Mild (C) Moderate (D) Severe

One is the Novel Angular Binary Pattern (NABP), and another approach is the Kernel-Based Convolutional Neural Networks (Kernel-Based CNN). This proposed practice attained an accuracy score of 97.39% [11].

Digital Camera Image

Digital Camera Images are captured and accessed using digital cameras like smartphone cameras. The camera is quite feasible and can be used smoothly in comparison to the fundus camera and slit-lamp device. Therefore, the usage of digital cameras has an enormous conceivable for cataract screening in the future and requires trained specialists and experienced ophthalmologists for cataract detection and classification (Fig. 6).

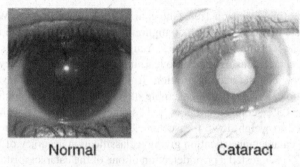

Normal   **Cataract**

Fig. 6. Digital camera images with cataract and no cataract

2.2 Machine Learning Techniques

This section depicts the machine learning techniques and algorithms used in the previous works of literature for cataract detection and classification. Machine Learning (ML) is a subfield of Artificial Intelligence (AI) that shows the capability of systems to perceive the given data by tracking down the best parameters and updating the weights within a general model using algorithms like K-Nearest Neighbor, Decision Tree, Logistic Regression, Support Vector Machines (SVM) and Random Forests. Over the past years, researchers have worked with various modernized machine-learning algorithms and methods for the detection and classification of cataracts lying on the severity level of cataracts. These techniques consist of 2 parts: extraction of features and features classification visible in Fig. 7. Table 1 summarizes all the machine learning algorithms or techniques for cataract detection and classification lying on distinct ophthalmic imaging modalities used in the previous works of literature.

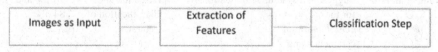

Fig. 7. A Figure showing a flowchart on machine learning lying on cataract classification/grading

The existing literature works with various machine learning methods and algorithms using distinct imaging modalities are given, like Li et al. [12, 13] proposed Support Vector Machine Regression (SVR) that classifies the hardness level of cataracts lying on slit-lamp images and attained fair classification outcomes. Xu et al. [14, 15] utilized the Group Sparsity Regression (GSR) method and Similarity Weighted Linear Reconstruction (SWLR) method for the classification of cataracts and gained great classification outcomes for the same. Yang et al. [16] use G filtering along with the Hat Transform to enrich fundus retinal images. After that, they removed the noise with the Trilateral filters, and a Backpropagation NN (BPNN) is worn for cataract classification. The accuracy score attained by this model is 82.9%. Zheng et al. [17] proposed a system that uses two-dimensional Discrete Fourier Transform configurations of fundus images to classify cataracts. This system used Principal Component Analysis (PCA) and the LDA for dimensionality reduction, and the classifier was trained and tested with the Adaboost algorithm. The score gained by this system is 81.52%. Fan et al. [18] utilized the wavelet method for cataract classification by making use of the algorithms like Random Forest, and Gradient Boosting Decision Tree. Guo et al. [19] proposed Discrete Wavelet Transform and Discrete Cosine Transform configurations to train and test the model for the detection of cataracts from fundus images. Yang et al. [20] suggested a system for the classification of cataracts using ensemble learning to attain good accuracy. This system captures an independent set of configurations from fundus images, that involves texture-lying configurations along with an ensemble of BPNN and SVM classifiers that gave the 93.2% and 84.5% scores. Qiao et al. [21] suggested a model in which a common algorithm is used to weigh down the vector elements with the use of a Support Vector Machine for cataract classification giving a classification accuracy of 87.52%. A.B. Jagadale et al. [22] suggested a prior detection of one of the cataract forms that occur in the middle of an eye with Slit-lamp imaging modality with an accuracy of 90.25%.

Table 1. Methods for cataract detection & classification lying on distinct imaging techniques

Authors	Methods	Imaging modality	Purpose	Accuracy
Yang et al. [16]	Use G filtering along with the Hat Transform to enrich fundus retinal images	Fundus Image	Cataract Classification	82.9%
Zheng et al. [17]	Use of two-dimensional Discrete Fourier transform configurations of fundus images to classify cataracts	Fundus Image	Classification of Cataracts	81.52%
Yang et al. [20]	Suggested a system for the classification of cataracts using ensemble learning	Fundus Image	Detection and Classification of Cataracts	93.2% and 84.5%
Qiao et al. [21]	Suggested a model in which a common algorithm is used to weigh down the vector elements	Fundus Image	Cataract Classification	87.52%
A.B. Jagadale et al. [22]	Suggested a prior detection of one of the cataract forms that occur in the middle of an eye	Slit-lamp Image	Cataract Detection	90.25%
Khan et al. [23]	The Support Vector Machine algorithm is conceived to classify cataracts	Digital Camera Image	Cataract Classification	89.26%

Apart from the machine learning methods and algorithms mentioned above, there are various other advanced ML methods introduced for the classification of cataracts, like Linear Discriminant Analysis, k-means clustering, Markov random field (MRF), and Bayesian network, etc. Despite having the various state-of-the-art ML algorithms, there are still several challenges as in the diagnosis of cataracts, extraction of tiny blood vessels and handcrafted features play an important role, which is not obvious with conventional ML methods. Therefore, in the past years, several research has been done by scholars using the deep learning approaches as it does not require human intervention and the automatic process happens to extract features from images.

2.3 Deep Learning Techniques

This section depicts the deep learning techniques and algorithms used in previous works of literature for cataract detection and classification. Deep Learning is a part of a wide family of Machine Learning. There are huge varieties of deep learning techniques ranging from Artificial Neural Networks (ANN), Back Propagation Neural Networks (BPNN), Multilayer Perceptron (MLP) Neural Networks, Classification Neural Networks (CNN), and Recurrent Neural Networks (RNN) to Transfer Learning (TL) based methods have been used for solving the tasks in various fields. In this survey, we mainly focus on cataract detection and classification task. Table 2 summarizes all the deep learning algorithms or methods for the detection and classification of cataracts lying on distinct ophthalmic imaging modalities used in the previous works of literature.

Table 2. Methods for cataract detection and classification lying on distinct imaging modalities

Authors	Methods	Imaging modality	Purpose	Accuracy
Caixinha et al. [25]	Cataract Classification using a multilayer perceptron model	Fundus Image	Classification of cataracts	96.7%
Zhang et al. [26]	Eight-layer DCNN is used along with a soft-max classifier	Fundus Image	Detection and Grading of cataracts with four class levels	93.5% and 86.6%
Ran et al. [27]	Proposed a system in which a Deep Convolutional Neural Network was discussed to classify the cataract into six levels of classes	Fundus Image	Grading of cataracts with six class levels	90.7%

(continued)

Table 2. (*continued*)

Authors	Methods	Imaging modality	Purpose	Accuracy
Li [28]	Worked with two distinct sorts of Convolutional Neural Networks, one is an 18-layered ResNet design and the other is a 50-layered ResNet design	Fundus Image	Cataract Detection and Classification	97.2% and 87.7%
Yadav et al. [29]	Proposed an ensemble learning-lying composite approach that comprises distinct prior trained convolutional neural networks along with the transfer Learning	Fundus Image	Cataract Classification	96.25%
Gao, Lin, & Wong [32]	An unsupervised convolutional-recursive neural networks (CRNN) method is used to automatically learn features for grading the severity of nuclear cataracts	Slit-lamp Image	Cataract Classification	88.4%
Qian, Patton, Swaney, Xing, & Zeng [33]	Use supervised training of CNN to identify different areas of cataracts in the lens	Slit-lamp Image	Cataract Classification	96.1%
Peterson et al. [34]	An iPhone X camera with an external device for the flashlight is used to capture the eye region images and evaluated to measure luminance reflection and colour features of the lens using a CNN	Digital Camera Image	Cataract Classification	98.2%

The existing literature works with various deep learning methods and algorithms using distinct imaging modalities are given like, Zhou et al. [24] proposed a shallow multilayer perceptron neural network model with the extraction of features for the classification of cataracts lying on the hardness levels. Caixinha et al. [25] also used the

Multilayer Perceptron Neural Network model to classify cataracts and achieved 96.7% of accuracy. Ran et al. [27] proposed a system in which a Deep Convolutional Neural Network was discussed to classify the cataract into six levels of classes. The score they got by this model for cataract classification with six levels is 90.8%. Li [28] worked with two distinct sorts of Convolutional Neural Networks, one is an 18-layered ResNet design and the other is a 50-layered ResNet design, used to determine the cataract to get the location area where cataracts are available. These systems attained an accuracy score of 97.2% and 87.7%. Yadav et al. [29] proposed an ensemble learning-lying composite approach that comprises distinct prior trained convolutional neural networks along with the transfer Learning to fragment configurations from fundus images for 4-class classification with an accuracy of 96.25%. Xiong et al. [30] have also worked with two sorts of configurations: texture configuration and other configurations. The score gained from this model for the six-level grading of cataracts is 91.5%. Imran et al. [31] proposed a composite system by adding a prior trained Convolutional Neural Network with the classifier to grade four-level classes of cataracts with a score of 95.6% on a huge fundus dataset. Gao, Lin, & Wong [32] proposed an unsupervised CRNN method to grasp configurations for grading the hardness of nuclear cataracts using a slit-lamp imaging modality with a score of 88.4%. Peterson et al. [34] proposed a method in which they used an iPhone X camera for cataract classification with an external device for the flashlight and maximum resolution. The eye region images were captured and evaluated to measure the shine replica and colour features of the lens using a Convolutional Neural Network. This method gained an accuracy of 98.2%.

Over the past years, researchers have put their effort into various modernized deep learning algorithms and techniques for cataract detection and classification. However, there are still several challenges and gaps that remain in the detection and classification of cataracts. First, the understandability of deep learning algorithms and techniques is not good as that of conventional machine learning algorithms and techniques. Second, the availability of large, labelled datasets along with quality images and manual marking is a singular and not a flaw-free method.

3 Conclusions and Recommendations

In recent years, Cataract Detection and Classification emerges as a vital and hot research area among scholars used to boost the quality of life for cataract patients. In this survey, we highlight the previous works of literature for cataract detection and classification using modernized machine learning and deep learning algorithms with distinct ophthalmic imaging modalities that achieve great accuracy. Besides that, there are still several challenges and gaps remain that medical systems face in meeting the ongoing eye care necessity of the world's community of citizens. Some of the gaps available in existing methods are as follows:

1. It is admired that none of the prior studies tested their distinctive algorithms further on the external dataset. As a result, the generalizability of these methods has not yet been demonstrated.
2. The practical usefulness of the discussed algorithms in real-world contexts (such as communities, eye hospitals, or primary care settings) also has to be assessed.

Therefore, the emergence of new algorithms or the polishing of existing algorithms, and the availability of huge, well-labeled clinical datasets remain a challenge. To alleviate the above-mentioned challenge, it must ensure that hospital medical records (clinical data and images) are grouped with secure cloud storage using seamless connections and provide high-quality ground truth data that can be further used for the creation of new algorithms so that the emergence of machine learning and deep learning techniques with distinct ophthalmic imaging modalities gives new opportunities to scholars in developing innovative systems and strategies in the fields of cataract detection and classification in the future.

4   Future Work

In previous studies, distinct ophthalmic imaging modalities such as fundus retinal, slit lamps, and digital images were used by the researchers with modernized machine learning and deep learning approaches. From them, slit lamp images need a medical gadget that is valuable and not portable and fundus cameras are also expensive and need expertise. Hence, the usage of digital images using a digital camera like a smartphone camera with the operation of machine learning and deep learning techniques can be the future of cataract screening and needs to be further explored by researchers in providing a friendly recommendation for ophthalmologists, especially in rural areas having limited healthcare facilities and infrastructure.

References

1. WHO. World Report on Vision: Executive summary (2019). https://www.who.int/docs/. Accessed 04 June 2021
2. Vashist, P., Senjam, S.S., Gupta, V., Gupta, N., Kumar, A.: Definition of blindness under national program for control of blindness: do we need to revise it? Indian J Ophthalmol. 65(2), 92–96 (2017). https://doi.org/10.4103/ijo.IJO_869_16. PMID: 28345562; PMCID: PMC5381306
3. Pathak, S., Raj, R., Singh, K., Verma, P.K., Kumar, B.: Development of portable and robust cataract detection and grading system by analyzing multiple texture features for Tele-Ophthalmology. Multimedia Tools Appl. 81(16), 23355–23371 (2022). https://doi.org/10.1007/s11042-022-12544-5
4. WHO. Global data on visual impairments (2012). https://www.who.int/blindness/. Accessed 04 June 2021
5. NPCBVI. National blindness and visual impairment survey India 2015-19: a summary report; (2020). https://npcbvi.gov.in/writeReadData/mainlinkFile/File341.pdf. Accessed 14 June 2021
6. Long, E., et al.: An artificial intelligence platform for the multihospital collaborative management of congenital cataracts. Nat. Biomed. Eng. 1(2), 1–8 (2017)
7. https://www.aao.org/eye-health/treatments/what-is-slit-lamp. Accessed 06 June 2022
8. Wikipedia contributors. Fundus photography. In Wikipedia, The Free Encyclopedia (2022). https://en.wikipedia.org/w/index.php?title=Fundus_photography&oldid=1083927539. Accessed 06 June 2022
9. Parikh, C.H., Fowler, S., Davis, R.: Cataract screening using telemedicine and digital fundus photography. Investig. Ophthalmol. Vis. Sci. 46(13), 1944 (2005)

10. Raju, B., Raju, N.S.D., Akkara, J.D., Pathengay, A.: Do it yourself smartphone fundus camera – DIYretCAM. Indian J. Ophthalmol. **64**(9), 663–667 (2016). https://doi.org/10.4103/0301-4738.194325

11. Sirajuddin, A., Balasubramanian, A., Karthikeyan, S.: Novel angular binary pattern (NABP) and kernel based convolutional neural networks classifiers for cataract detection. Multimedia Tools Appl. (2021). https://doi.org/10.1007/s11042-022-13092-8

12. Li, H., et al.: An automatic diagnosis system of nuclear cataract using slit-lamp images. In: 2009 Annual International Conference of the IEEE Engineering in Medicine and Biology Society, Minneapolis, MN, USA, pp. 3693–3696. IEEE (2009). https://doi.org/10.1109/IEMBS.2009.5334735

13. Li, H., et al.: A computer-aided diagnosis system of nuclear cataract. IEEE Trans. Biomed. Eng. **57**(7), 1690–1698 (2010)

14. Xu, Y., et al.: Automatic grading of nuclear cataracts from slit-lamp lens images using group sparsity regression. In: Mori, K., Sakuma, I., Sato, Y., Barillot, C., Navab, N. (eds.) MICCAI 2013. LNCS, vol. 8150, pp. 468–475. Springer, Heidelberg (2013). https://doi.org/10.1007/978-3-642-40763-5_58

15. Xu, Y., Duan, L., Wong, D.W.K., Wong, T.Y., Liu, J.: Semantic reconstruction-based nuclear cataract grading from slit-lamp lens images. In: Ourselin, S., Joskowicz, L., Sabuncu, M.R., Unal, G., Wells, W. (eds.) MICCAI 2016. LNCS, vol. 9902, pp. 458–466. Springer, Cham (2016). https://doi.org/10.1007/978-3-319-46726-9_53

16. Yang, M., Yang, J.J., Zhang, Q., Niu, Y., Li, J.: Classification of retinal image for automatic cataract detection. In: Proceedings of the 2013 IEEE 15th International Conference on e-Health Networking, Applications & Services (Healthcom 2013), Lisbon, pp. 674–679 (2013). https://doi.org/10.1109/HealthCom.2013.6720761

17. Zheng, J., Guo, L., Peng, L., Li, J., Yang, J., Liang, Q.: Fundus image-based cataract classification. In: Proceedings of the IEEE International Conference on Imaging Systems and Techniques (IST), Santorini, Greece, pp. 90–94 (2014). https://doi.org/10.1109/IST.2014.6958452

18. Fan, W., Shen, R., Zhang, Q., Yang, J.J., Li, J.: Principal component analysis-based cataract grading and classification. In: Proceedings of the 17th International Conference on E-Health Networking, Application & Services (HealthCom), Boston, MA, pp. 459–462. IEEE (2015). https://doi.org/10.1109/HealthCom.2015.7454545

19. Guo, L., Yang, J.J., Peng, L., Li, J., Liang, Q.A.: Computer-aided healthcare system for cataract classification and grading based on fundus image analysis. Comput. Ind. **69**, 72–80 (2015). https://doi.org/10.1016/j.compind.2014.09.005

20. Yang, J.J., et al.: Exploiting ensemble learning for automatic cataract detection and grading. Comput. Methods Programs Biomed. **124**, 45–57 (2016). https://doi.org/10.1016/j.cmpb.2015.10.007

21. Qiao, Z., Zhang, Q., Dong, Y., Yang, J.J.: Application of SVM based on genetic algorithm in classification of cataract fundus images. In: Proceedings of the 2017 IEEE International Conference on Imaging Systems and Techniques (IST), Beijing, China. IEEE, pp. 1–5 (2017). https://doi.org/10.1109/IST.2017.8261541

22. Jagadale, A.B., Sonavane, S.S., Jadav, D.V.: Computer aided system for early detection of nuclear cataract using circle hough transform. In: Proceedings of the 2019 3rd International Conference on Trends in Electronics and Informatics (ICOEI), Piscataway, NJ, USA, vol. 2019, pp. 1009–1012. IEEE (2019)

23. Khan, A.A., Akram, M.U., Tariq, A., Tahir, F., Wazir, K.: Automated computer aided detection of cataract. In: Abraham, A., Haqiq, A., Ella Hassanien, A., Snasel, V., Alimi, A.M. (eds.) AECIA 2016. AISC, vol. 565, pp. 340–349. Springer, Cham (2018). https://doi.org/10.1007/978-3-319-60834-1_34

24. Zhou, Y., Li, G., Li, H.: Automatic cataract classification using deep neural network with discrete state transition. IEEE Trans. Med. Imaging **39**(2), 436–446 (2019)
25. Caixinha, M., Jesus, D.A., Velte, E., Santos, M.J., Santos, J.B.: Using ultrasound backscattering signals and nakagami statistical distribution to assess regional cataract hardness. IEEE Trans. Biomed. Eng. **61**(12), 2921–2929 (2014)
26. Zhang, L., Li, J., Han, H., Liu, B., Yang, J., Wang, Q.: Automatic cataract detection and grading using deep convolutional neural network. In: Proceedings of the 2017 IEEE 14th International Conference on Networking, Sensing and Control (ICNSC), Calabria, Italy, pp. 60–65. IEEE (2017). https://doi.org/10.1109/ICNSC.2017.8000068
27. Ran, J., Niu, K., He, Z., Zhang, H., Song, H.: Cataract detection and grading based on combination of deep convolutional neural network and random forests. In: 2018 International Conference on Network Infrastructure and Digital Content (IC-NIDC), Guiyang, China. IEEE, pp. 155–159 (2018). https://doi.org/10.1109/ICNIDC.2018.8525852
28. Li, J., et al.: Automatic cataract diagnosis by image-based interpretability. In: Proceedings of the 2018 IEEE International Conference on Systems, Man, and Cybernetics (SMC), Miyazaki, Japan, pp. 3964–3969 (2019). https://doi.org/10.1109/SMC.2018.00672
29. Yadav, J.K.P.S., Yadav, S.: Computer-aided diagnosis of cataract severity using retinal fundus images and deep learning. Comput. Intell. **38**(4), 1450–1473 (2022). https://doi.org/10.1111/coin.12518
30. Xiong, Y., He, Z., Niu, K., Zhang, H., Song, H.: Automatic cataract classification based on multi-feature fusion and SVM. In: Proceedings of the 2018 IEEE 4th International Conference on Computer and Communications (ICCC), Chengdu, China, pp. 1557–1561. IEEE (2018). https://doi.org/10.1109/CompComm.2018.8780617
31. Imran, A., Li, J., Pei, Y., Akhtar, F., Yang, J.J., Dang, Y.: Automated identification of cataract severity using retinal fundus images. Comput. Methods Biomech. Biomed. Eng. Imaging Vis. **8**(6), 691–698 (2020). https://doi.org/10.1080/21681163.2020.1806733
32. Gao, X., Lin, S., Wong, T.Y.: Automatic feature learning to grade nuclear cataracts based on deep learning. IEEE Trans. Biomed. Eng. **62**(11), 2693–2701 (2015)
33. Qian, X., Patton, E.W., Swaney, J., Xing, Q., Zeng, T.: Machine learning on cataracts classification using squeeze net. In: Proceedings of the 2018 4th International Conference on Universal Village (UV), Piscataway, NJ, USA, vol. 2, pp. 1–3. IEEE (2018)
34. Peterson, D., Ho, P., Chong, J.: Detecting cataract using smartphone. Invest. Ophthalmol. Vis. Sci. **61**(7), 474 (2020)

Automatic Identification of Cataract by Analyzing Fundus Images Using VGG19 Model

Rakesh Kumar[1], Vatsala Anand[1], Sheifali Gupta[1(✉)], Maria Ganzha[2,3], and Marcin Paprzycki[3,4]

[1] Chitkara University Institute of Engineering and Technology, Chitkara University, Rajpura, Punjab, India
{rakesh.ece,vatsala.anand,sheifali.gupta}@chitkara.edu.in
[2] Warsaw University of Technology, Warsaw, Poland
maria.ganzha@pw.edu.pl
[3] Systems Research Institute Polish Academy of Sciences, Warsaw, Poland
marcin.paprzycki@ibspan.waw.pl
[4] Warsaw Management University, Warsaw, Poland

Abstract. Nowadays, cataracts are one of the prevalent eye conditions that may lead to vision loss. Precise and prompt recognition of the cataract is the best method to prevent/treat it in early stages. Artificial intelligence-based cataract detection systems have been considered in multiple studies. There, different deep learning algorithms have been used to recognize the disease. In this context, it has been established that the training time of the VGG19 model is very low, when compared to other Convolutional Neural Networks. Hence, in this research, the VGG19 model, for automatic cataract identification in fundus images, has been proposed for healthy lives. The performance of the VGG19 is explored with four different optimizers, i.e. Adam, AdaDelta, SGD and AdaGrad and tested on a collection of 5000 fundus images. Overall, the best experimental results reached 98% precision of classification.

Keywords: Cataract · Biomedical · VGG19 · Transfer learning · Classification · Convolutional Neural Network · Disease

1 Introduction

One of the widespread eye conditions, the cataract, is also one of the most common causes of vision loss and blindness. Cataract is an eye condition, where the lens becomes foggy or cloudy, which results in blurred image. Hence, a person suffering from the cataract is not able to see clearly [2, 3]. Moreover, the blurred vision causes problems with seeing both at night and in bright light. Globally, many factors like different lifestyle, old age, gender, profession, financial status, personal care, traditions, and conventions, cause eye diseases. According to available research, comparing the population in tropical and temperate countries, it was found that the chronic eye infections in tropical populations are very often due to the elements such as dust, humidity, sunshine, and other environmental conditions [4]. Moreover, people over the age of 60 are more likely to get the cataract [5]. Injury to the lens, or retina, can also cause the cataract. While some people get the cataract due to genetic diseases, any chronic eye disease, surgery, or diabetes can also cause cataracts [6]. Long-term use of steroids has also been implicated as a potential cause the cataract. There's no certain way to prevent, or to slow, the development of the cataract, but some measures to keep eyes healthy have been suggested (see, for instance, [7, 8]). Blindness, caused by the cataract, can be prevented with early detection and prompt treatment. However, quite often, people with vision problems shy away from getting their eyes checked [9]. Here, it should be noted that, since the cataract develops slowly, its initial effects are difficult to detect. Moreover, the change in the vision depends on the location and size of the cataract [10]. After the diagnosis of the cataract, the typical treatment is surgical [11]. In this context, the contribution of this work is to explore the performance of the VGG19 deep learning model, applied to the automatic identification of the cataract in the fundus images. Dataset of 5000 images is being used for model training and performance testing. To improve model performance, four different optimizers (Adam, AdaDelta, SGD and AdaGrad) have been experimented with. Overall, the best obtained results reached 98% accuracy.

The remaining parts of this work are organized as follows. Review of pertinent literature is presented in Sect. 2. Rationale of the proposed approach to cataract detection, using VGG19 model, is outlined in Sect. 3. Next, Sect. 4 describes the experimental setup, including the dataset used in the experiments. Analysis of obtained results is presented in Sect. 5, while Sect. 6 concludes the paper and outlines future research directions.

2 Literature Survey

Modern automatic cataract detection algorithms utilize various machine learning and deep learning algorithms. In this section, leading works devoted to both approaches are summarized.

Jing Ran, et al. [1] proposed combining deep convolutional neural network (DCNN) with Random Forest (RF) for cataract grading. Using RF and delivering six-level cataract grading schema provides more accurate results than use of DCNN and a four-level differentiation. This, in turn, should help doctors to better understand patient's condition. Overall, the best-reported accuracy was 90.69%. Tao Li, et al. [2] summarized recent developments in deep learning for fundus image classification. Additionally, 33 publicly available datasets have been described. Xiaohangwu et al. [3] proposed a three-step method, consisting of (1) cataract detection, (2) cataract diagnosis, and (3) establishing cataract etiology and severity. The reported performance, for the cataract detection (step one of the approach), was 99.8%. In the second step, 99.2% accuracy was obtained for the cataract diagnosis. Finally, 99.93% accuracy was reached for the mydriatic slit lamp mode, and 99.71% for non-mydriatic slit lamp mode. Linglin Zhang et al. [4] used deep convolutional neural network and experimented with the dataset that consisted of 5620 images, obtained from multiple hospitals. The standard DCNN architecture was complemented with pool5 layer to "explain meaning" of extracted features. The reported accuracy was 93.52%. Xiaoqing Zhang et al. [6] analyzed performance of two methods, named deep learning and conventional technique. Moreover, various challenges in automatic grad-based cataract detection were discussed. On the basis of an overview of results, found in the literature, it was suggested that, currently, deep learning-based methods lead to the best accuracy (outperform other machine learning approaches).

In this context, Md. Rajib Hossain, et al. [7] proposed the cataract detection system based on deep convolutional neural network. Here, the model consists of four blocks with convolutional and max-pooling layers. Next, flatten+dense layer is applied, followed by three dropout+dense layers. Here, the first two use 128: ReLu and 256: ReLu functions, while the last one uses the sigmoid function. The reported accuracy was at 95.77% on test sets. Wanga, et al. [8] explored use of U-Net for convolutional decoder technique that provides accurate segmentation, using many colors, of fundus images from the optic disc (OD) area. This, in turn, helps to detect glaucoma. Here, unique subnets and decoding convolutional blocks have been applied. They increased the contrast and improved reliability and accuracy during segmentation of the OD area. In the best-reported results, the accuracy was approximately 97%. Yanyan Dong, et al. [9] preprocessed the fundus images using the maximum entropy method. Next, Caffe-based deep learning network was applied, to automatically extract distinctive features from the fundus images. Finally, automatically extracted features have been classified using softmax algorithms. The best obtained accuracy was 94.82%. Duoru Lin, et al. [10] proposed a technique to identify hereditary cataract for high-risk babies. Their goal was to develop a practical model to identify infants at high risk for congenital cataracts (CC), a leading cause of preventable childhood blindness. The CC identification model accurately distinguished between the CC patients and the healthy children. It has been envisioned as a supplementary screening tool. Using the CC screening, based solely on AI analyses, the CC identification model distinguished between the CC patients and the healthy children with 95% accuracy and has been envisioned as a supplementary screening tool. Sakshi et al. [11] proposed Handwritten Mathematical Symbols and Expressions (HMSE) model to help in feature extraction, pattern recognition, computer vision and artificial intelligence area. They implemented up to 65% accurately in their technique's studies. Amitoj Singh et al. [12] guided for data organization in software engineering and management and judged through each dimension.

3   Cataract Detection Using VGG19 Model

The VGG19 is a new model, based on very deep convolutional network that provides image recognition with better accuracy on large-scale image datasets. It represents an attempt to use in practice the idea of transfer learning. In particular, VGG19 has been pre-trained on an image dataset, with focus on edges, as well as horizontal and vertical lines. This pre-trained model is then trained on the image dataset pertinent to the task at hand. For recognizing the cataract, existence of the pre-trained model should be reducing the training time, support extraction of features, existing in the dataset, with better accuracy. This was also the main reason for the decision to explore usability of the VGG19 to the cataract detection, based on automatic analysis of the fundus images [13]. As the name implies, the VGG19 has nineteen convolution layers, complemented by one SoftMax layer, three fully connected layers, and five MaxPool layers. Note that the VGG19 architecture is based on CNN models with 3x3 filters [14–16]. The pertained VGG19 model, used for finding the cataract, was the VGG19 model available from the Kaggle platform.

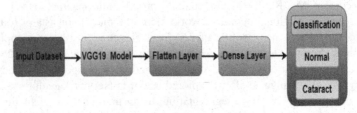

Fig. 1. VGG19 architecture

Figure 1 shows the VGG19 architecture, with Flatten and Dense layers, that has been applied to the cataract detection task. Here, to convert a single long continuous linear vector into 2-dimensional arrays from the pooled feature maps, the Flatten Layer is used. The Dense Layer collects information from neurons from the previous layer and performs the final classification task (deciding that the image represents either a Normal eye or an eye with the Cataract).

```
Model: "sequential"

Layer (type)                Output Shape              Param #
=================================================================
vgg19 (Functional)          (None, 7, 7, 512)         20024384

flatten (Flatten)           (None, 25088)             0

dense (Dense)               (None, 1)                 25089
=================================================================
Total params: 20,049,473
Trainable params: 25,089
Non-trainable params: 20,024,384
```

Fig. 2. VGG19 model [17]

Figure 2 specifies the VGG19 model that was available in the Kaggle repository. The functional VGG19 model used 2,00,24,384 parameters, while the dense layer had 25089 parameters. The total number of parameters was 20,049,473, with 25089 being trainable and 20,024,384 non-trainable.

4 Input Dataset

The Ocular Disease Intelligent Recognition (ODIR; [18]) is the most important (and popular) dataset, and a set of resources, for detecting eye diseases, among these that are (currently) available in Kaggle. The structural ophthalmic database, in ODIR, contains information about 5,000 individuals, including age, color images of the fundus in both eyes, and diagnostic terms used by the experts. Note that this is the standard way that fundus images, from ODIR, have been used across the literature (see, for instance, work reported in [19]). Here, Fig. 3 shows the sample images of the Cataract and the Normal eye, identified by the VGG19 model.

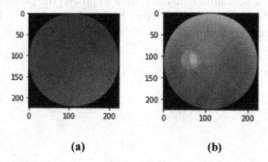

(a)   **(b)**

Fig. 3. Dataset sample (a) Cataract (b) Normal [17]

It should be stressed that in the ODIR dataset, actual patient data that has been collected by the Shang Gong Medical Technology Co., Ltd. This data has been collected from multiple hospitals and medical facilities in China. Note that these institutions used variety of cameras, from manufactures such as Canon, Zeiss, and Kowa, to capture high-resolution fundus images. This introduces additional dimension of heterogeneity, which has not been explored (as it was out of scope of this work). In order to ensure quality, experts have been involved to split (annotate) the data into two classes (normal and cataract).

5   Experimental Results

While using the training data from the dataset, the training accuracy can be improved by application of optimizers. Therefore, together with the VGG19 model, four optimizers: SGD, AdaGrad, AdaDelta, and Adam have been tried. Here, it should be stated that, in each case, the performance was better than that of the VGG19 without an optimizer. Therefore, it has been decided that, for sake of space preservation, only results with optimizers will be reported. Performance has been measured using standard accuracy, loss, as well as precision, recall and F1 performance measures. Moreover, in each case, for completeness of results, the confusion matrix has been reported. Since large number of approaches has been applied to the cataract recognition problem, and various datasets have been used for model training (see, Sect. 2), it is difficult to unequivocally establish, which of them should be treated as the baseline. Moreover, various approaches, reported in the literature, applied different hyperparameter tuning, while in work reported here only state-of-the-art optimizers have been applied to boost performance (no other tuning has been explored). Therefore, only results of experiments with the VGG19 model, combined with the four optimized, are reported. However, it can be easily noticed that they are competitive with these reported in state-of-the-art literature, reported in Sect. 2.

5.1 Results for the SGD Optimizer

Let us start reporting results from these obtained for the simplest optimizer, the Stochastic Gradient Descent (SGD), which has been applied to all training parameters. In the SGD optimizer, a sigmoid activation function is used to predict the probability distribution. The obtained results have been summarized in Fig. 4. Specifically, Fig. 4(a) and (b) represent model accuracy and model loss, for training and validation, respectively, for 15 training epochs.

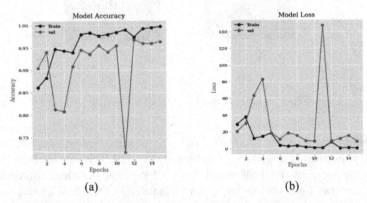

(a)   (b)

Fig. 4. Results when SGD optimizer was applied; (a) accuracy (b) loss.

Accuracy and Loss Analysis for the SGD Optimizer. First, let us note that after 15 epochs no further improvement has been observed and the training process has been stopped. Second, Fig. 4(a) shows the training accuracy value of 0.9471 has been reached during the 3rd epoch, followed by 0.9793 for the 6th epoch. After 15 epochs, training accuracy reached 0.9966. Next, the value of validation accuracy has also been systematically improving from the 3rd epoch (0.8119) to the 15th epoch (0.9633). Moreover, it can be also seen, in Fig. 4(b) that, as expected, the value of loss is decreasing (for both training and validation) as the training progresses, and reaches 8.1597 for the validation loss after 15 epochs.

Confusion Matrix Parameters Analysis. The confusion matrix is often used to evaluate (and visualize) the effectiveness of classification models [20, 21]. The confusion matrix obtained in the experiments with the SGD optimizer is represented in Fig. 5. Here, 127 true negative images and 83 true positive images have been reported. Additionally, application of trained model resulted in 2 false negatives and 6 false positives.

Finally, in Table 1, the class '0' represent 'cataract' whereas the class '1' represents non-cataract. Here, the precision, recall, F1-score values are reported for experiments with the SGD optimizer. Moreover, the overall accuracy resulting from application of the SGD optimizer to the trained VGG19 model was 0.96.

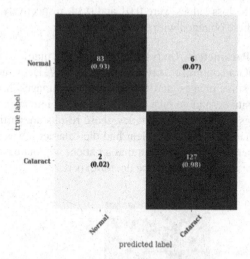

Fig. 5. Confusion matrix for SGD optimizer

5.2 Results for the AdaGrad Optimizer

Adaptive gradient algorithm (AdaGrad) works based on the gradient optimization. Ada-Grad adapts data component-wise, by adding information resulting from the training process. Accuracy and loss (for training and validation), obtained when VGG19 model has been combined with the AdaGrad optimizer, have been reported in Fig. 6.

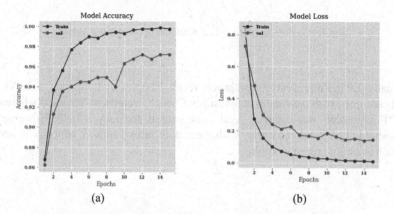

Fig. 6. Results when AdaGrad optimizer was applied; (a) accuracy (b) loss

Accuracy and Loss Analysis for the AdaGrad Optimizer. Figure 6(a) and (b) depict the training accuracy (0.95) and loss (0.15) after 3 epochs, respectively. The accuracy of training and validation improved with each training epoch. Similarly, model loss declined with each subsequent training epoch. In both cases, the plateau has been reached around the 15th epoch. There, accuracy for training and validation were 0.99 and 0.97 respectively, while the loss values were 0.01 and 0.14, respectively. These results are, clearly, better than these obtained with the SGD optimizer.

Confusion Matrix Parameters Analysis. The confusion matrix obtained in the experiments with the AdaGrad optimizer is represented in Fig. 7. Here, the model accurately classified 92 true positive images, and 120 true negative images. Moreover, the model predicted 6 false positives, and no false negatives. Taking into account that the considered problem belongs to medical informatics, these results are particularly good. This is because in no situation where the patient had the cataract it was not detected. False positives, on the other hand, would mean that a patient without cataract would be sent for further diagnostics; and cleared by the doctor/expert.

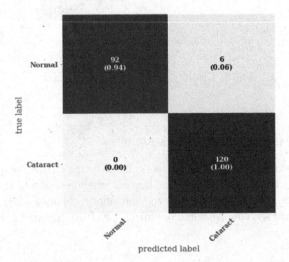

Fig. 7. Confusion matrix when AdaGrad optimizer was applied

Finally, Table 1 represents the precision, recall, F1-score and accuracy of the VGG19 model combined with the AdaGrad optimizer. Class '0' represent 'cataract' whereas the class '1' represents non-cataract. Again, the overall accuracy is 0.96. However, this results should be considered jointly with the lack of false positives, which points to this approach as being more favorable.

Table 1. Confusion matrix parameters for AdaGrad optimizer.

Class	Precision	Recall	F1-score	Accuracy
0	0.98	1.00	0.94	0.96
1	0.98	0.95	1.00	
Average value	0.98	0.975	0.97	

5.3   Results for the AdaDelta Optimizer

The AdaDelta optimizer uses a gradient descent stochastic algorithm that is based on adaptive learning rates applied throughout the training. The results obtained when the AdaDelta has been applied to the VGG19 model have been summarized in Fig. 8.

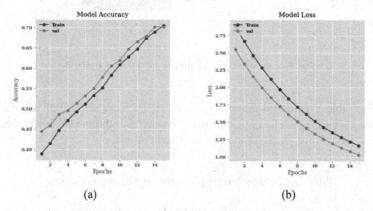

(a)   (b)

Fig. 8. Results when AdaDelta optimizer was applied; (a) accuracy (b) loss

Accuracy and Loss Analysis for the AdaDelta Optimizer. Figure 8(a) and (b) represent the training accuracy and loss values, for training and validation. The accuracy of the results improved with each epoch, while the model loss declines. For the 15th epoch, accuracy for training and validation are 0.70 and 1.16, while the loss values re 1.16 and 0.03, respectively. Since the remaining optimizers reach accuracy of more than 96% at 15 epochs or less we have decided to stop training at 15 epochs also in this case. This decision was also supported by the fact that one of main reasons for using VGG19 model was to reduce training time/cost.

Confusion matrix parameters analysis, For the model trained using AdaDelta optimizer, Fig. 9 represents the confusion matrix. Here, 63 true positives and 90 true negatives are reported. Moreover, 33 false negatives and 32 false positives have been obtained. Obviously, these results could have been better, if the training continued. However, capturing the snapshot of accuracy at 15th epoch (when 3 out of 4 systems have reached already reached their "peak performance") turned out to be one of interesting aspects of the experimental comparison.

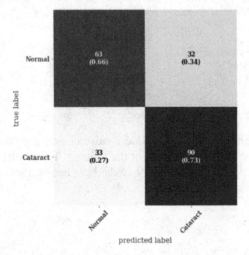

Fig. 9. Confusion matrix for the AdaDelta optimizer

Finally, Table 2 represents that precision, recall, F1 score and overall accuracy. The overall accuracy of AdaDelta optimizer is 0.70, which is substantially lower than these reported in Table 1, for the SGD and the AdaGrad optimizer.

Table 2. Confusion matrix parameter on Adadelta optimizer.

Class	Precision	Recall	F1-score	Accuracy
0	0.66	0.66	0.66	0.70
1	0.74	0.73	0.73	
Average value	0.70	0.695	0.695	

5.4 Results for the Adam Optimizer

Finally, as the last one, Adam optimizer has been experimented with. It applies a binary cross entropy loss function [20], to predict the probability of an output and uses this prediction to optimize model parameters. Results obtained when the VGG19 model was combined with Adam optimizer, for 15 training epochs, have been presented in Fig. 10.

Accuracy and Loss Analysis for the Adam Optimizer. For the 3rd epoch, the training accuracy and loss are around 0.98 and 0.10, respectively, while the validation accuracy

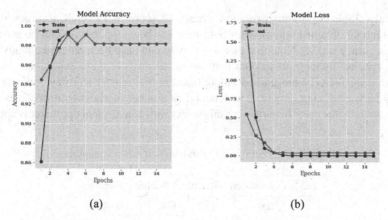

Fig. 10. Results for the Adam optimizer; (a) accuracy, (b) loss.

and loss are approximately 0.97 and 0.17, respectively. Starting from epoch 6, both accuracy and loss stabilize. Moreover, they remain "more stable" than what can be seen in the cases of SGD and AdaGrad optimizers. Here, the final accuracy of validation is about 0.98, while the loss is approximately 0.03.

Confusion Matrix Parameters Analysis. In Fig. 11 the confusion matrix for the results of the VGG19 model combined with Adam optimizer are presented. The model identified 110 true negatives and 104 true positives. Moreover, 3 false positives and 1 false negative results have been delivered. The last one is slightly worrisome, as this would mean that a single patient with the cataract would not be spotted.

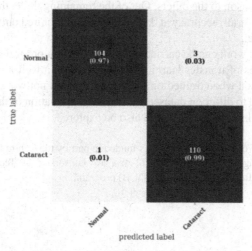

Fig. 11. Confusion matrix for VGG19 with Adam optimizer

Finally, Table 3 presents precision, recall, F1-score and overall accuracy of the VGG19 model obtained when Adam optimizer has been applied. As can be seen, the overall accuracy is 0.98. Note that this accuracy would have been reached already after about 6 epochs. Taking into account that one of the goals of these experiments was to establish if the VGG19 pre-trained model (used in the transfer learning context) does actually reduce training cots (e.g. understood as number of epochs), it becomes clear that it is actually the case. In particular, this works well when Adam optimizer is applied.

Table 3. Confusion matrix parameter on Adam Optimizer.

Class	Precision	Recall	F1-score	Accuracy
0	0.99	0.97	0.98	0.98
1	0.97	0.99	0.98	
Average value	0.98	0.98	0.98	

6 Concluding Remarks

In this work, the pre-trained (for image recognition) VGG19 deep neural network has been applied to the task of cataract detection in fundus images. Moreover, the performance of the VGG19 has been explored when four different optimizers have been applied, i.e., SGD, AdaGrad, AdaDelta, and Adam. Out of these, when evaluated from the point of view of training effort, the performance of the AdaDelta optimizer turned to be inferior in comparison to the others. Out of the remaining three, the Adam optimizer delivered the best, overall, accuracy at 0.98. Moreover, it required only 6 training epochs to reach stable model.

In the future, the work can be carried out using different pre-trained models, further exploring boundaries of transfer learning. Moreover, it is a well-known fact that very often models obtained when trained on a given data set do not work well "in real life" and/or when applied to different datasets. Therefore, application of the trained model to large variety of available datasets should also be explored.

Acknowledgment. Work of Maria Ganzha is funded in part by the Centre for Priority Research Area Artificial Intelligence and Robotics of Warsaw University of Technology within the Excellence Initiative: Research University (IDUB) program.

References

1. Ran, J., Niu, K., He, Z., Zhang, H., Song, H.: Cataract detection and grading based on combination of deep convolutional neural network and random forests. In: 2018 International Conference on Network Infrastructure and Digital Content (IC-NIDC), pp. 155–159. IEEE, August 2018

2. Li, T., et al.: Applications of deep learning in fundus images: a review. Med. Image Anal. **69**, 101971 (2021)
3. Wu, X., et al.: Universal artificial intelligence platform for collaborative management of cataracts. Br. J. Ophthalmol. **103**(11), 1553–1560 (2019)
4. Zhang, L., Li, J., Han, H., Liu, B., Yang, J., Wang, Q.: Automatic cataract detection and grading using deep convolutional neural network. In: 2017 IEEE 14th International Conference on Networking, Sensing and Control, pp. 60–65. IEEE, May 2017
5. Oda, M., Yamaguchi, T., Fukuoka, H., Ueno, Y., Mori, K.: Automated eye disease classification method from anterior eye image using anatomical structure focused image classification technique. In: Medical Imaging 2020: Computer-Aided Diagnosis, vol. 11314, pp. 991–996. SPIE, March 2020
6. Zhang, X.Q., Hu, Y., Xiao, Z.J., Fang, J.S., Higashita, R., Liu, J.: Machine learning for cataract classification/grading on ophthalmic imaging modalities: a survey. Mach. Intell. Res. **19**(3), 184–208 (2022). https://doi.org/10.1007/s11633-022-1329-0
7. Hossain, M.R., Afroze, S., Siddique, N., Hoque, M.M.: Automatic detection of eye cataract using deep convolution neural networks (DCNNs). In: 2020 IEEE Region 10 Symposium (TENSYMP), pp. 1333–1338. IEEE, June 2020
8. Wang, L., et al.: Automated segmentation of the optic disc from fundus images using an asymmetric deep learning network. Pattern Recogn. **112**, 107810 (2021)
9. Dong, Y., Zhang, Q., Qiao, Z., Yang, J.J.: Classification of cataract fundus image based on deep learning. In: 2017 IEEE International Conference on Imaging Systems and Techniques (IST), pp. 1–5. IEEE, October 2017
10. Lin, D., et al.: A practical model for the identification of congenital cataracts using machine learning. EBioMedicine **51**, 102621 (2020)
11. Kukreja, V.: A retrospective study on handwritten mathematical symbols and expressions: Classification and recognition. Eng. Appl. Artif. Intell. **103**, 104292 (2021)
12. Singh, A., Kukreja, V., Kumar, M.: An empirical study to design an effective agile knowledge management framework. Multimed. Tools Appl. **82**, 12191–12209 (2023). https://doi.org/10.1007/s11042-022-13871-3
13. Junayed, M.S., Islam, M.B., Sadeghzadeh, A., Rahman, S.: CataractNet: an automated cataract detection system using deep learning for fundus images. IEEE Access **9**, 128799–128808 (2021)
14. Chalakkal, R.J., Abdulla, W.H., Thulaseedharan, S.S.: Quality and content analysis of fundus images using deep learning. Comput. Biol. Med. **108**, 317–331 (2019)
15. Anand, V., Gupta, S., Nayak, S.R., Koundal, D., Prakash, D., Verma, K.D.: An automated deep learning models for classification of skin disease using Dermoscopy images: a comprehensive study. Multimedia Tools and Applications **81**(26), 37379–37401 (2022). https://doi.org/10.1007/s11042-021-11628-y
16. Anand, V., Gupta, S., Altameem, A., Nayak, S.R., Poonia, R.C., Saudagar, A.K.J.: An enhanced transfer learning based classification for diagnosis of skin cancer. Diagnostics **12**(7), 1628 (2022)
17. Aloysius, N., Geetha, M.: A review on deep convolutional neural networks. In: 2017 International Conference on Communication and Signal Processing (ICCSP), pp. 0588–0592. IEEE, April 2017
18. Anand, V., Gupta, S., Koundal, D., Nayak, S.R., Nayak, J., Vimal, S.: Multi-class skin disease classification using transfer learning model. Int. J. Artif. Intell. Tools **31**(02), 2250029 (2022)
19. Khan, M.S., et al.: Deep learning for ocular disease recognition: an inner-class balance. Comput. Intell. Neurosci. **2022**, 1–12 (2022)

20. Son, J., Shin, J.Y., Kim, H.D., Jung, K.H., Park, K.H., Park, S.J.: Development and validation of deep learning models for screening multiple abnormal findings in retinal fundus images. Ophthalmology **127**(1), 85–94 (2020)
21. Li, Y., Duan, P.: Research on the innovation of protecting intangible cultural heritage in the "internet plus" era. Procedia Comput. Sci. **154**, 20–25 (2019)

Sensors Based Advanced Bluetooth Pulse Oximeter System

Jaspinder Kaur[✉], Ajay Kumar Sharma, and Divya Punia

Department of Computer Science and Engineering, National Institute of Technology, Delhi,
New Delhi, India
jaspinderkaur@nitdelhi.ac.in

Abstract. Arduino based Bluetooth-equipped pulse oximeter is a measurement device that uses near infrared spectroscopy to measure blood pressure, and is designed with the HC-05 Bluetooth module. It can be employed using a smart mobile application or hardware. The oximeter uses a I2C 16 * 2 display module, which is a parallel data converter chip that works seamlessly with the LCD display module. This chip can convert the I2C data into parallel data, which is required by the LCD display. The portable terminal uses a digital algorithm to determine the value of the oxygen saturation and the pulse rate, and it does so through the smart mobile app interface. The designed oximeter can help doctors to keep a time to time check on the patient's pulse and Spo2 level from anywhere in the hospital via their mobile phones, which would especially be helpful to keep the doctors, nurses distant from the patients during any Pandemic. The paper presents a novel model of Arduino based Bluetooth Pulse Oximeter using sensors and Bluetooth module with its applications in various sectors.

Keywords: Oximeter · Sensor · Bluetooth · Arduino NANO

1 Introduction

Oximeters were designed as a noninvasive way to measure arterial oxygen saturation in the blood of a patient undergoing medical treatment. Clinical treatment for critically sick patients, patient monitoring under anesthesia in procedures, research of breath status while sleeping, and so on are some of the applications of pulse oximetry [1, 2]. The goal of this project was to see if a (Bluetooth-capable) pulse oximeter with the same functionality as a regular wired oximeter could be developed to make the medical process of measuring and monitoring oxygen saturation in the bloodstream more comfortable and convenient for the patient receiving medical care [3, 4]. Basic connectivity of components for integrated circuits and microcontrollers were adapted and implemented in order to create a functional Bluetooth pulse oximeter. The raw sensor data, as well as the oxygen saturation and heart rate, were effectively shown in digital form when this gadget was connected to a cell phone through Bluetooth. Future implementations for the project's development are also mentioned, as well as enhancements that may be done for a more efficient model. To determine an individual's pulse oximetry, the effectiveness of this

S. Sachdeva et al. (Eds.): BDA 2022, LNCS 13830, pp. 149–160, 2023.
https://doi.org/10.1007/978-3-031-28350-5_12

transport mechanism and total system performance are assessed. In order to determine a person's SPo2, several characteristics of oxygenated hemoglobin and deoxygenated hemoglobin are needed. As a result of our study, we want to provide assistance to the country by doing the best that we can within the given time frame. With the right resources, time management, and collective efforts, it was a novel Bluetooth based pulse oximeter model [5, 6].

2 Design and Implementation

One of the biggest problems with the Bluetooth Oximeter system is that it's not easy to teach every user how to use it. Main objectives include designing and utilizing a product that can assist the people of the nation in the spread of a pandemic using artificial intelligence (AI) capable of making this feasible, and using it via Bluetooth on your mobile phones. The system is extremely portable due to its tiny size, yet it accomplishes its function well. The Bluetooth module is used to make the connection. In addition to lowering the deployment costs, this will also make it easier to upgrade and reconfigure the system in the future. Arduino Nano and Max30100 Pulse Oximeter show Heart Rate and Blood Oxygen on 162 I2C compatible LCD Module and also communicate the blood oxygen, heart rate, or Pulse rate information to the android application built in Android Studio using the wireless Bluetooth technology. Blood Oxygen Concentration (BOC) is measured in percentage whereas the heart rate or pulse rate is measured in BPM, commonly known as beats per minute. To monitor Blood Oxygen and BPM readings, you may use an android mobile phone application if you are near the circuit [7, 8]. The key components involved in the designing of the proposed system are explained below.

1. Arduino NANO
   The ATmega328-based Arduino Nano is a tiny, comprehensive, and breadboard-friendly board (Arduino Nano 3.x). It offers a lot of the same features as the Arduino Duemilanove, but it comes in a different packaging.

Fig. 1. Arduino NANO supposedly smaller than Arduino UNO

It just has a DC power connector and uses a Mini-B USB cable rather than a normal one. In comparison to other, larger Arduino boards, the Arduino Nano boards a seemingly little but important benefit. It will fit on a breadboard since it does not have the same uneven pin spacing as the original Arduino designs (which was supposedly a mistake in the original design file). Figure 1 shows the Arduino Nano board [9].

2. Bluetooth module

HC-05 Bluetooth module is a simple 6 pin configuration VCC, GND, transmitter, receiver, key, and LED. Bluetooth module which is shown in Fig. 2. It is a module that transmits data from one device to another without wires (wirelessly) using the phenomenon called radio frequency transmission. The principal purpose of this device is to return the serial communication that previously used to be connected through wires now to be wireless. Bluetooth contains two types of devices, called transmitter and receiver. This HC-05 module is set by default baud rate of 9600 bps but can be changed and made between 1200 bps to 1.35 Mbps [2, 10]. Here, the master is using the Bluetooth module but still the module itself is being used as a slave to take the inputs from the mobile devices and hence this device takes the input wirelessly from the user's smartphone and performs the task by which the follower called "slave" follows this "master" device.

Fig. 2. HC-05 bluetooth module

3. I2C 16 * 2 LCD Display

The I2C communication interface, this 16 * 2 Arduino LCD screen uses I2C. It indicates that the LCD display only requires four pins: VCC, GND, SDA, and SCL. As many as four digital/analog pins will be saved. All connections are standard XH2 stud-type XH2 [11].

Figure 3 shows the infrared sensors. It is a 3-pin configuration sensor. The pins are VCC, GND, OUT, respectively. It runs on a 5 V DC power source and is able to work in the range of 100 mm. Two I2C interfaces are required, which may be linked through Dupont Line or an I2C dedicated cable.

Fig. 3. Infrared sensors, used for direction handling, based on reverse logic activation on specific sides of the rover.

4. Female Headers

A pin header (sometimes known as a "header") is a type of electrical connection. Though there are various name variants of male and female connections, the female counterparts are commonly referred to as a female socket header as shown in Fig. 4.

Fig. 4. Female header

5. Voltage regulator

A voltage regulator is a circuit that produces and maintains a constant output voltage regardless of input voltage or load circumstances. The voltages from a power source are kept within a range that is compatible with the other electrical components by voltage regulators (VRs). Figure 5 depicts the voltage regulator.

Fig. 5. Voltage regulator

6. **PCB Board**

As shown in Fig. 6, a printed circuit board (PCB) uses conductive rails, pads, and other features carved from one or more sheet layers of copper bonded onto and/or between sheet layers of a non-conductive substrate to physically support and electrically link electrical or electronic components.

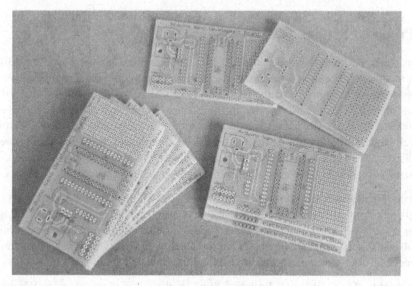

Fig. 6. PCB board

3 Applications

This simple yet complex oximeter can be used in various fields of operations such as in hospitals, army camps, schools/university, especially on COVID patients. As discussed above the applications for these types of oximeters are vast, they can be modified at major scale as per the user's needs [12, 13].

Some major sectors for application are explained as follows:

Hospital – During this ongoing Pandemic, as we all know it was utmost necessary to maintain physical distance between each other, in such situations it became really difficult for Doctors/nurses as they were the most prone to it because of their regular checkups to various Covid-19 patients. To keep a constant check on their patient health report they had to enter their wards, which also lead to a sudden spike in COVID cases in terms of doctors. Thus our Bluetooth Pulse Oximeter, would have been a perfect replacement for wire oximeters, to keep a constant check on their BPM and SPo2 level even from far distance on their respective mobile phones.

Army Camps – During Extrinsic workout, the trainers could also keep a check on the BPM levels of the army pursuits and watch their performance from a closer approach.

Schools/Universities – The Bluetooth oximeters are definitely the most successful in the education sector as it was the first thought and application. It is easy to make with correct knowledge and guidance provided by teachers. As per the new CBSE rules Coding has been made compulsory for students onward 6th class. Hence it would be a Project of application for them as well they can use it in the school dispensary.

4 Proposed System

In this paper, Arduino based Bluetooth Pulse Oximeter model using sensors and Bluetooth module have been presented and discussed in detail with various applications. In this system, we have proposed the HC05 Bluetooth Module, PCB Board, Max 30100 Sensor and 16 * 2 I2C Display, due to these main components this project has been possible, as the Hc05 Module gives us connectivity from the oximeter to the I2C display or even your android mobile devices, the PCB board holds all the connection between the display to Bluetooth module to the sensor, coming to Max 30100 sensor, we used this sensor as it doesn't give out radiation. Following are the reasons for using bluetooth.

Current Healthcare Situation - During this COVID-19 phase, there has been a lot of insufficient supplies of medical equipments in both rural and urban hospitals. In villages Some doctors still used old techniques to find the pulse rate of a patient by putting a thumb on the nerves and reading beats per minute, this can be a very serious problem as chances of a doctor or nurse himself/herself getting infected are more [14].

Convenient - Quick check of Bmp and Spo2 level per sec monitoring via Bluetooth on cellphone. Thus a safe distance would be maintained between the patient and doctor/relatives as well. As per during this pandemic period, a much needed product would add as another safety equipment for patients and would somehow help to break the chain further.

Memory Recordings - O2 and heart rate data may be recorded and/or streamed using wireless pulse oximeters with the necessary recording app or Windows software. The Masimo Mightysat and the Nonin Connect both offer the ability to transmit data to an app that tracks O2 and heart rate.

Safety - These days' doctors suggest people to stay at home than come to hospital unless very necessary. Thus the Covid-19 positive patients who are at home quarantine, are asked to send their oxy level 5 different times a day so that the doctors can evaluate any progress in the patient and guide accordingly to the family members or patient as well [15].

5   Architecture and Methodology

The global epidemic of COVID-19 demonstrated how people must keep their distance from one another in every scenario, not just for their own protection but also for the safety of others. The project concept was designed with the same circumstances in mind that everyone is working in these days. The project revolves around the Arduino NANO, a popular and frequently used development board. We picked this board for a variety of reasons, including how easy it is to work on, how well-built the libraries are, how much data is accessible, and so on. We began the design process with a straightforward yet practical approach. We made Arduino based Bluetooth Oximeter in such a way that it can be modified in any way or form with just a little effort and which also makes it one of the unique oximeters that are capable of these kinds of manipulations, that even students of a school or university with accurate knowledge and teacher's guidance.

The main objective behind this project was to design and fabricate a patient monitoring heart rate signals. The motto was to develop wireless system of monitoring system using Bluetooth Module. We develop a data monitoring system using integration between IC and Mobile platforms and to provide real time data at 'HOME' setting.

The sensors that we used were the best possible that we could get in the budget and in the difficult situations due to the global pandemic. All the components that we used such as Arduino NANO, Female Headers, HC-05 Bluetooth module, IC 16 * 2 LCD Display, Voltage Regulator, PCB Board, (diameter-7.5 cm), (width-1.5 cm), jumper wires, were the best we could find in the market.

This is the circuit schematic as it has been changed. The Arduino Nano is powered by an LM7805 voltage regulator-based 5 V regulated power supply. The DC female power connector is J1. This is where a 12 V adapter, battery, or solar panel is attached. Ensure that the power supply voltage does not exceed the LM7805 voltage regulator's input voltage limit. The Arduino Nano's Vin pin is then linked to the controlled voltage. Also, connect the power supply's GND to the Arduino board's ground pin.

The connection between the Arduino and the max30100 pulse Oximeter sensor stays the same.

The I2c supported the 16 * 2 LCD this time. The 16 * 2 LCD's SDA and SCL pins are linked to the Arduino's analogue pins A4 and A5. The VCC and GND pins of the LCD I2C Converter are linked to 5 V and GND, respectively.

The HC05 Bluetooth module's RX (Receiver) and TX (Transmitter) pins are linked to Arduino pins 2 and 3. The Bluetooth module's GND and +5 V pins are linked to the Arduino's GND and 5 V. Figure 7 shows the Arduino NANO development board.

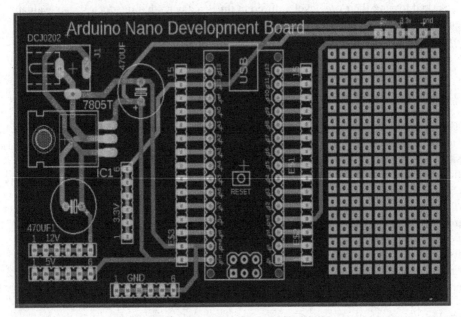

Fig. 7. Arduino NANO development board

Moreover, PCB for the Arduino Nano board is created, which would serve as the development board. For the 3.3 V, 12 V, 5 V, and ground, female headers are added. The VeroBoard on the right side can be used to solder additional electrical components. Moreover, PCB for the Arduino Nano board is created, which would serve as the development board. For the 3.3 V, 12 V, 5 V, and ground, female headers are added. The VeroBoard on the right side can be used to solder additional electrical components. On the left and right sides of the Arduino Nano, female headers are added for connecting the jumper wires. All of the connections were double-checked. Figure 8. Depicts the connections of Max30100 Pulse Oximeter, Bluetooth, and 16 * 2 LCD Display with Arduino NANO.

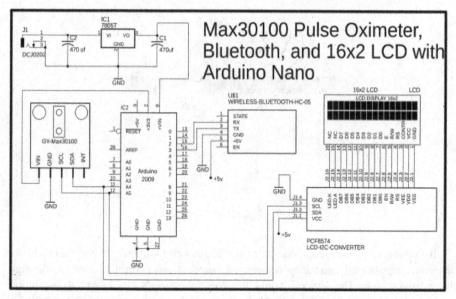

Fig. 8. Connections of Max30100 Pulse Oximeter, Bluetooth, and 16 * 2 LCD display with Arduino NANO

6 Results and Discussion

Coding was done on the Arduino IDE platform. The codes were developed in such a manner that the outputs of the Oximeter can be easily modified with not much effort. As the Arduino based Bluetooth Oximeters have to be used on different people, thus the coding has to be modified according to the needs as all the different persons these oximeters will be detecting will have a decent or wide range of different BPM and SPO2 levels.

We used the same program as before with a few minor changes. We added the Liquid Crystal library that supports I2C and created pins according to the circuit design. I included instructions for delivering the BPM and Oxygen Percentage values to the android mobile phone application, as well as instructions for displaying the BPM and Oxygen percentage information on the I2C compatible 16 * 2 LCD module in the void loop function. The GY-Max30100 Pulse Oximeter includes a total of 5 male headers that are clearly labelled as Vin, GND, SCL, SDA, and INT, as you can see. This is an i2c-compatible sensor that connects with the Arduino board through the I2C bus. So then we started placing the components and finished the soldering job.

After every one of the associations are done, the Bluetooth is supposed to be turned On and connected with the I2c LCD display then maintain a good distance between the finger and the sensor for accurate readings. It should be noted that thumb should not be pressed tightly as it affects the blood O2 rate. After that, you could see the digital output both on the LCD screen as well on the cell phone connected via Bluetooth. Figure 9 displays the system block diagram.

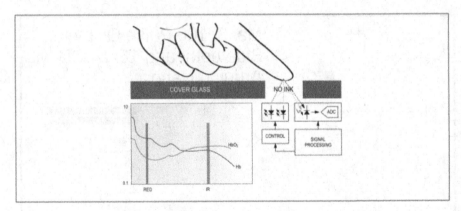

Fig. 9. System block diagram

By connecting basic components to the Bluetooth Oximeter, it was theoretically proved to operate well, and the appliances were successfully controlled remotely through the android mobile. The device not only monitors the patient's data, heart rate, and blood pressure, but it also tells if the patient's finger is off the oximeter or not on correctly.

Fig. 10. Final model design of Bluetooth Pulse Oximeter

The heart rate, often known as the pulse rate or heartbeat, is measured in BPM (beats per minute), whereas the Blood Oxygen Concentration is recorded in % age (percentage). Though MAX30100 is a very smart sensor. But sometimes it may show wrong readings if the finger is not placed correctly on the sensor. Figure 10 depicts the final model design of Bluetooth Pulse Oximeter.

The Arduino based Pulse Oximeter assembly was made successfully with the help of all the members of the team and the desired output was achieved. We were able to maneuver the Bluetooth Pulse Oximeter according to our needs. SpO2 and BPM levels were also verified in real life scenarios. Table 1 shows the cost of production of the final Bluetooth Pulse Oximeter.

Table 1. Cost of production of final Bluetooth Pulse Oximeter.

S.no	Name of equipment	Quantity	Price (rupees)
1	Arduino NANO	1	420
2	Female headers	10	250
3	Voltage regulator	1	50
4	PCB Board	1	150
5	MAX30100	1	160
6	HC-05	1	300
7	I2C 16 * 2 LCD	1	175
8	12 V battery	2	700
9	Jumper wires	3 sets	300
		Total	2505

7 Conclusion

In this paper, the authors have presented a novel model of Arduino board based Bluetooth Pulse Oximeter using sensors and Bluetooth module, formed from a thought to a real-life SpO2 and BPM level detecting system. This system is definitely capable of being presented at higher levels where authors can present it as a startup. This model will be helpful in various medical fields as well as an education purpose, as per the current scenario it can lead to a very well built startup in the medical sector for patients suffering through COVID-19 and thus will keep the hospital staff safe as well. Moreover, easy to use interface and low cost make the proposed model attractive for its use in schools and universities as well.

8 Future Scope

Wearable fitness trackers are all the rage right now. While most devices just track basic activity levels, an increasing number of gadgets, such as Apple's upcoming iWatch, are adding pulse and calorie burn data to the list of metrics. While there are a variety of pulse oximeters on the market that can measure both blood oxygen saturation and pulse, they aren't as wearable as a Fitbit. These gadgets are generally attached to the fingers or earlobe and employ stiff traditional circuitry. There is a wide scale of improvement in our model as it can be modified into wearable gadgets as well where any person feeling High Blood pressure or Low Blood Pressure can send the SpO2 and BPM levels to any doctor within his link from a distance. The authors can also work on the body's outer cover for protection and attractive looks and better sensors can be utilized in the coming future.

References

1. Communication Engineering: Pulse oximeter with Bluetooth interface. Int. J. Electr. Electron. Data Commun. **2**(6), 100–103 (2014)
2. Minaie, A., Sanati-Mehrizy, R., Paredes, L.E., Morris, J.: Design of a Bluetooth-enabled wireless pulse oximeter. In: ASEE Annual Conference & Exposition (2019)
3. Morón, M., Luque, R., Casilari, E.: On the capability of smartphones to perform as communication gateways in medical wireless personal area networks. Sensors **4**(1), 575–594 (2014)
4. Dai, Y., Luo, J.: Design of noninvasive pulse oximeter based on Bluetooth 4.0 BLE. In: Proceedings - 2014 7th International Symposium on Computational Intelligence and Design, vol. 1, no. 3, pp. 100–103 (2014)
5. Szakacs-Simon, P., Moraru, S., Perniu, L.: Pulse oximeter based monitoring system for people at risk. In: CINTI 2012 - 13th IEEE International Symposium on Computational Intelligence and Informatics, Proceedings, pp. 415–419 (2012)
6. Aithal, A.: Wireless Sensor Platform for Pulse Oximetry (2015)
7. Pak, G., Park, K.: Advanced pulse oximetry system for remote monitoring and management. J. Biomed. Biotechnol. (2012)
8. Moron, M., Casilari, E., Luque, R., Gázquez, J.: A wireless monitoring system for pulse-oximetry sensors. In: Proceedings - 2005 Systems Communications, ICW 2005, Wireless - ICHSN 2005, High Speed Networks - ICMCS 2005, vol. 2005, pp. 79–84 (2005)
9. Louis, L.: Working principle of Arduino and using it as a tool for study and research. Int. J. Control Autom. Commun. Syst. **1**(2), 21–29 (2016)
10. Rathod, K., Vatti, P., Nandre, M., Yenare, S.: Smart door security using Arduino and Bluetooth application. Int. J. Curr. Eng. Sci. Res. **4**, 73–77 (2017)
11. Arunpradeep, N., Niranjana, G., Suseela, G.: Smart healthcare monitoring system using IoT. Int. J. Adv. Sci. Technol. **29**(6), 2788–2796 (2020)
12. Shreya, S., Chatterjee, K., Singh, A.: A smart secure healthcare monitoring system with Internet of Medical Things. Comput. Electr. Eng. **101**, 107969 (2022)
13. Motwani, A., Shukla, P., Pawar, M.: Ubiquitous and smart healthcare monitoring frameworks based on machine learning: a comprehensive review. Artif. Intell. Med. **134**, 102431 (2022)
14. Sangeetha, T., Kumutha, D., Bharathi, D., Surendran, R.: Smart mattress integrated with pressure sensor and IoT functions for sleep apnea detection. Measur. Sens. **24**, 100450 (2022)
15. Rahman, M., Hossain, G., Challoo, R., Rizkalla, M.: iRestroom: a smart restroom cyberinfrastructure for elderly people. Internet Things **19**, 100573 (2022)

Information Interchange of Web Data Resources

Saral Anuyojan: An Interactive Querying Interface for EHR

Kanika Soni[1]([✉]), Shelly Sachdeva[1], Arpit Goyal[1], Aryan Gupta[1], Divyanshu Bose[1], and Subhash Bhalla[2]

[1] National Institute of Technology Delhi, Delhi, India
{kanikasoni,shellysachdeva,191210012,191210014,
191210019}@nitdelhi.ac.in
[2] University of Aizu, Aizuwakamatsu, Japan

Abstract. Maintaining a lifelong medical record is impossible without proper standards. For an individual, different records from different sources must be brought meaningfully together for them to be of some use. To achieve this, we need a set of pre-defined standards for information capture, storage, retrieval, exchange, and analytics. It has been found that electronic health records can enhance the quality and safety of care while improving the management of health information and clinical data. While electronic health records have so much potential, it is difficult to use them. It requires queries written in AQL to interact with the EHR database. Writing AQL queries is a complex as well as a tedious task. An interface is needed that can speed up the querying process thereby enhancing efficiency. Considering the importance of using electronic health records and the difficulty of using them, we aim to design Saral Anuyojan which is a system that consists of components such as user interface, query translator, and interface manager. The user interface takes input from the user and then the query translator converts it into AQL queries for further processing. Then the AQL query is sent to the backend (EHRbase) which stores it in a standard format. Finally, the output is returned as a visual interpretation on the user interface. Requirements of the clinicians and patients are limited (view and update) so they can be implemented without much complexity. Since, EHR has a complex structure difficult to use by a non-technical user, this problem is re-solved in our approach by making an easy user user-interface we are bypassing the long and complex approach of learning AQL. The proposed system Saral Anuyojan helps in improving the management and efficiency of the healthcare sector.

Keywords: User interface · Electronic Health Record (EHR) · Healthcare · Databases

1 Introduction

A collection of different medical records created during any clinical interaction or incident is called an Electronic Health Record (EHR) [1]. Nowadays, the development of self-care, homecare equipment, and the ongoing generation of useful healthcare data can be quite helpful in a variety of applications. But a lifetime medical record is just not feasible without guidelines. A person's many records from various sources need to be usefully combined. Images, clinical codes, and data must all be captured, stored, retrieved, exchanged, and analyzed according to a set of predetermined standards in order to accomplish this. The Personal Health Record Management System, known as MyHealthRecord [2], was created by the Ministry of Health and Family Welfare in conjunction with the Ministry of Electronics and IT, Government of India [3] for Indian citizens in response to the growing trend toward a healthcare system focused on citizens as opposed to one focused on institutions. The Ministry of Health & Family Welfare, Government of India, released the Electronic Health Record (EHR) Standards for India in September 2013 [4] and revised them in 2016. The collection of standards provided there was picked from among the best standards for electronic health records that were available and in use globally, weighing their applicability to and usefulness for India. Different criteria that must be applied for various jobs are explicitly specified. The "Framework for Mobile Governance 2012" [5] of the Department of Electronics and IT, Ministry of Communication & Information Technology, Government of India, shall be relevant with regard to the specific design and use of such applications on mobile devices.

Need for EHR: The Digital India [6] effort has helped to accelerate the number of healthcare services being digitized in India's various healthcare delivery organizations. In addition to better administration of health information and clinical data, practitioners and physicians have asserted that the use of electronic health records can improve the quality and safety of healthcare. Along with greater patient-provider communication, electronic health records also make clinical data more portable, which aids public health professionals in understanding illness patterns and improving disease diagnosis. It is very beneficial for patients as well because it gives them access to better services and cuts down on unnecessary clinical tests.

1.1 OpenEHR

OpenEHR [7] is a virtual community working on means of turning health data from the physical form into electronic form and ensuring universal interoperability among all forms of electronic data. The primary focus of its endeavor is on electronic health records (EHR) and related systems. The essential outcome of the openEHR approach is systems and tools for computing health information at a semantic level, thus enabling true analytic functions like decision support, and research querying.

1.2 Archetype

Semantic modeling is provided through archetypes [9] (independent of software). These are outlined as restrictions on data whose occurrences follow a research model (RM). Language is not a barrier to the creation or translation of archetypes. Information structures and formal terminology are linked through archetypes. The archetypes impose restrictions on the acceptable structures of cases of the general class that belongs to

RM. EHRs use archetypes to support multiple content kinds in order to function as an information exchange platform. There will be many different ways to represent the EHR data. It is possible for the content to be unstructured, structured, semi-structured, or a combination of all three. These could be sentences, measured amounts (values and units), paragraphs, etc.

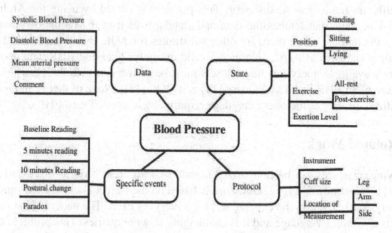

Fig. 1. Archetype for blood pressure

The blood pressure archetype, a clinical notion with a variety of diverse features, is depicted in Figure 1 [10]. We can see that while merging them under one archetype, the various pressures have independent representations, and the various postures, etc. need to be represented individually. The accuracy of each clinical notion is evaluated. An archetype on blood pressure measurement, for example, might contain a detailed and formal description of every-thing a clinician needs to know about a measurement, including how to interpret it in a way that is clinically safe. An archetype logically might be a specialization of another archetype or another archetype itself. So, they are adaptable and have many forms. They are flexible in terms of scope, all-purpose, and reusable. In contrast to data, archetypes are stored in repositories. The repository at any given location will always contain archetypes from libraries.

1.3 Archetype Definition Language

A formal language called Archetype Definition Language (ADL) [11] is used to express archetypes. Each ADL archetype is created while taking into account a reference model. Templates, which are defined locally, are used to apply archetypes to data. Extensible Markup Language (XML) [12] instance documents have previously been used to express archetypes. Instances of the parse tree that have been serialized are equal to XML archetypes. The Archetype Definition Language uses serialized form to represent archetypes. The cornerstone of ADL, an abstract language that also incorporates nomenclature, are frame logic queries, also referred to as F-logic. Any archetype's semantics can be rendered guarantee losslessly using an ADL archetype. ADL is a text-based language used to formally establish limitations on data instances of an RM.

1.4   Archetype Query Language

EHR base is queried using Archetype Query Language (AQL) as the results need to be standardized. The data stored in the EHRbase is heterogeneous and cannot be displayed as it is. The results need to be standardized before output. Also, the data cannot be requested directly as the format in which it is stored is not known and the formats of two different fields may be different. This problem is solved by using the Archetype Query Language that fetches the data and standardizes it in the required format and returns the output. While there are other substitutes for AQL, they are not as effective and comprehensive as AQL. To create a valid query for querying EHR data using the currently available query languages, users must be conversant with the persistent data structure of an EHR (such as XQuery [14] and OQL [15]). None of them can therefore be utilized directly as the query language required by integrated care EHRs.

2   Related Work

Querying interfaces can be categorized mainly in 4 different ways as discussed in the paper [17] as shown in Fig. 2. Our work is based on the 'direct manipulation' querying category. Table 1 shows the existing work in Querying EHR. The paper [18] states that AQL is a complex language and it is challenging to write queries. This problem can be solved by a user-friendly interface that can take the user input and script the corresponding AQL queries. The scope of queries can also be limited to only fetch and update. This would not only be a user-friendly method but would also reduce the chance of error in writing the queries. The paper [16] describes that the structure of an EHR is extremely complicated and may contain information on 100–200 parameters, each of which may have its values. To overcome this, the structure of the EHR base, the backend for interoperable clinical applications, can be limited to satisfy only the user requirements and limit the parameters for the specific use. The paper [17] identifies that joins are necessary for efficient implementation and maintaining the structure of the database. Implementing joins is very complex and cannot be done easily by a non-technical user like a clinician. To implement joins while maintaining the database structure and usability of the application, the designed interface must be interactive and interpretable. The different output requirements should be well-defined from the beginning so that corresponding joins can be implemented. In the paper [19] EHR stores heterogeneous data, and since we know UI is much better than raw data, so by creating an interface we generate a paper-free solution that increases the quality of software by structuring data. Key Principles which are taken care of during structuring are data accuracy and validity. During the structuring of data join statements are used which are complex to understand even for a skilled user. The paper [21] mentions the importance of three principles of interoperability, usability, and knowledge evolution. Interoperability would further help in cost reduction in the medical field and improve the treatment of the respective population. It permits the addition of new information without creating any schema. The paper [22] discusses the benefits of integrated architecture. It proposes an approach for integration that enables access to clinical data from different types of sources. Generates a collection of data from different sources, providing a large number of results, but the directories generated are too long when writing queries which makes it difficult for a user to understand. The paper [20] discusses about query language is essential for health information systems. The proposed study aims to develop a graphical interface for querying EHR data. The motive of the research is to meet the querying needs of healthcare consumers. It makes it easy to perform complex functions like joins and understand relationships between the entities in a complex EHR.

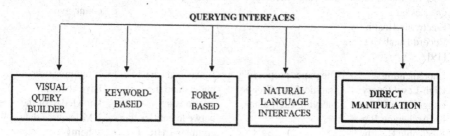

Fig. 2. Different categories of querying interfaces

Table 1. Existing work in EHR Querying along with its pros and cons

Title of the paper	Summary	Pros	Cons
Relational-like Query Language for Archetypes in Standardized Electronic Health Records Databases [16]	The paper identifies the need to develop a web-based app for efficient querying	• Compatible with a variety of databases (RDBMS, XML, NoSQL) • Workflow is easy • Provides high-level support	• Do not support multi-concept and nested queries • Requires skilled users
Integration through mapping - an OpenEHR-based approach for research-oriented integration of Health Information Systems [22]	The paper proposes an approach for integration that enables access to clinical data from different types of sources	• Allows the system to work on a centralized database • Possible to obtain results in a semi-structured format	• Directories are an issue for query creation since they are too long
AQBE - QBE Style Queries for Archetype Data [18]	The paper proposes an advanced DBMS architecture for repository services for health records and a step-by-step graphical query interface to allow semi-skilled users to write queries	• The user does not need any prior knowledge of archetype • It provides an easy interface view	• Error handling is much more difficult • Features like filtering and sorting are not implemented in AQBE

(*continued*)

Table 1. (*continued*)

Title of the paper	Summary	Pros	Cons
Dynamic Generation of Archetype-Based User Interfaces for Queries on Electronic Health Record Databases [19]	Provides an interface that improves the querying efficiency	• Solves the problem of semantic heterogeneity	• Complex queries pose issues in statements such as count, avg
Implementing High-Level Query Language Interfaces for Archetype-Based Electronic Health Records Database [20]	The paper proposed a graphical interface for querying EHR data	• It makes it easy to understand the relationship between entities in a complex EHR	• It requires users to have absolute knowledge of the XML source data schema
Evolving large-scale healthcare applications using open standards [21]	Discusses the importance of interoperability, usability, and knowledge evolution. Interoperability would aim to provide benefits on reusability and cost reduction which in turn help in medical advancement	• Dynamic adaptation to any regional language • It supports the modification of existing archetypes for developing countries	• Lacks multilingual support which affects Usability

3 Proposed Methodology

Problem Statement: Considering the utility of Electronic Health Records and the importance of making them accessible and easy to use, there is a need to design user-friendly interfaces for efficient querying. The structure of EHRs is highly complex as shown in Fig. 3. Since querying EHRs directly is a quite tedious task hence, the user who can be either a clinician or a patient needs interactive interfaces to easily query the EHR. There is a crucial demand of these interfaces to act as a mediator between the user and an EHR without the user requiring to know the entire structure of EHRs as a pre-requisite to querying.

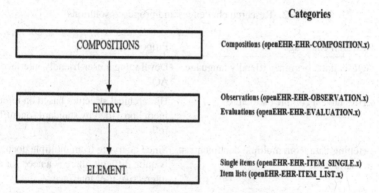

Fig. 3. Hierarchical structure of EHR and categories of openEHR archetypes

Research Challenges: Table 2 shows the various research problems that we may face during the implementation and the solutions that we propose to implement in Saral Anuyojan. The above-mentioned challenges have been figured out from the various research papers related to our work. Every research challenge has been mentioned with its corresponding proposed solution. The proposed solution has been derived based on the availability of tools and technology presently and the solutions can be implemented using tools and technologies that are currently available.

3.1 Proposed Methodology

Saral Anuyojan is a system that consists of components such as user interface, query translator, and interface manager. The user interface takes input from the user and then the query translator converts it into AQL queries for further processing. Then the AQL query is sent to the backend (EHRbase) which stores it in a standard format. Finally, the output is returned as a visual interpretation on the user interface. Requirements of the clinicians and patients are limited (view and update) so they can be implemented without much complexity. Required checks and filters will be applied for different users. As discussed earlier, EHR has a complex structure difficult to use by a non-technical user. This problem is resolved in our approach. By making an easy user user-interface we are bypassing the long and complex approach of learning AQL and understanding the structure of the dataset. In our approach, we are also using document embedding concept to resolve the challenge of complex querying from multiple documents.

Table 2. Research challenges and proposed solutions

S. No.	Research challenge	Proposed solution
1	AQL is not a beginner-friendly language	Developing a user-friendly interface for AQL
2	EHRs have a complex structure	Hierarchical structure based on reference model provided by standard (openEHR) is followed
3	Fetching data from multiple documents at the same time	Since querying from multiple documents is quite difficult, there is a need for an interactive user interface
4	Dealing with heterogeneous data stored in EHR	The data needs to be stored in the form of key-value pairs where value could belong to the various categories of data

3.2 Architecture

Figure 4 shows the architecture of Saral Anuyojan with all the components. The process flow has been described and the components are grouped into certain blocks for ease of reference. The main components as shown in the diagram are divided into three main phases.

Phase 1: Query Translator. The input is the query provided by the user and the output is either the re-quested information or an error message. The input is passed to the Query Translator which has three components. The parser takes in the input and parses it. Then the input keywords are mapped to the cor-responding keywords of AQL and the translation is completed. If there is some error in translation, then it is passed directly to the output. Otherwise, the query is sent to the database to perform the required functionality.

Phase 2: EHR Database. The AQL query is sent to the EHRbase [34] which is an open-source backend for clinical application systems and electronic health records. It is based on the openEHR specifications, an open platform architecture for creating adaptable and interoperable eHealth systems. The result of the AQL query is outputted as a preliminary result. EHRbase stores the heterogeneous Electronic Health Records in a standard format and returns results based on the input AQL query.

Phase 3: Interface Manager. The preliminary result data obtained from the Clinical Data Repository is converted into JSON [35] by a Document Extractor. The JSON data is filtered and formatted according to the output requirements and then displayed.

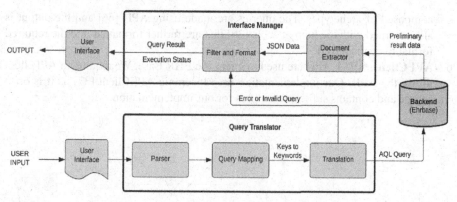

Fig. 4. Architecture of proposed Saral Anuyojan

4 Implementation and Results

4.1 Hardware and Software Requirements

CPU: 2.8 GHz Pentium 4, Pentium D 2.66 GHz, AMD Athlon 64 3600+ or better;
CPU SPEED: 3.0 GHz or greater; RAM: 2 GB or greater; OS: Windows 7/10/11/Linux/Mac; Google Chrome, Mozilla Firefox or equivalent browser; Internet connectivity to navigate.

4.2 Implementation Diagram

Figure 5 shows the implementation flow for Saral Anuyojan.

1. **Clinical Concepts Repository:** OpenEHR Clinical Knowledge Manager (CKM) [23] was used to fetch the archetypes that formed the base of the software. CKM serves as the repository for the archetypes that define the various clinical concepts. The required archetypes can be fetched from the OpenEHR CKM and can be used. It is an open-source repository.
2. **Template Designer:** The archetypes fetched from the Clinical Concepts Repository are used to create templates using OpenEHR Archetype Designer [25].
3. **Composition Designer like Cabolabs:** The templates created using the Archetype Designer were used to create compositions using a composition designer like Cabolabs [30]. Compositions are necessary to create the EHR base and for performing operations on it.
4. **Rest APIs:** The APIs were used to post the requests and templates to the database and get the re-quired output. The queries are sent to the EHR database with the help of APIs. The templates and compositions are also posted to the database using these APIs.
5. **Clinical Data Repository:** EHR Base has been used as the clinical data repository for our pur-pose. It stores Electronic Health Record (EHR) data and can be queried using AQL (Archetype Query Language). It is designed using templates and compositions

composed of archetypes. The queries are made using APIs [36] and the output is also returned with the help of APIs [36] that are further formatted into the required format to be displayed.

6. **API Client:** REST APIs are used to query the EHR Base. We require an API client to use the APIs. Our implementation uses Insomnia API Client [37] as it is open source and contains sufficient features for our implementation.

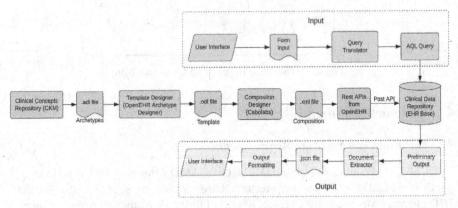

Fig. 5. Implementation diagram for Saral Anuyojan

4.3 Testbed

The following Table 3 shows some sample AQL queries of different categories. These queries were executed on our proposed interface "SARAL ANUYOJAN" and the corresponding results have been shown in Figs. 6, 7, 8 and 9.

Table 3. Sample queries

Query No.	Query	Result
Q1	Retrieving observations from compositions	Fig. 6
Q2	Retrieving doctor name from observations	Fig. 7
Q3	Retrieving systolic blood pressure greater than a value	Fig. 8
Q4	Retrieving diastolic blood pressure greater than a value and order by time	Fig. 9

4.4   Results

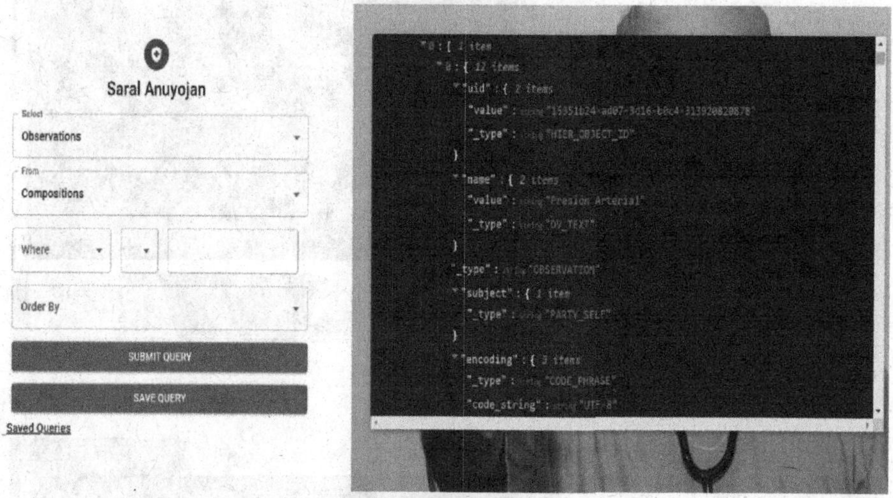

Fig. 6. Result for Q1

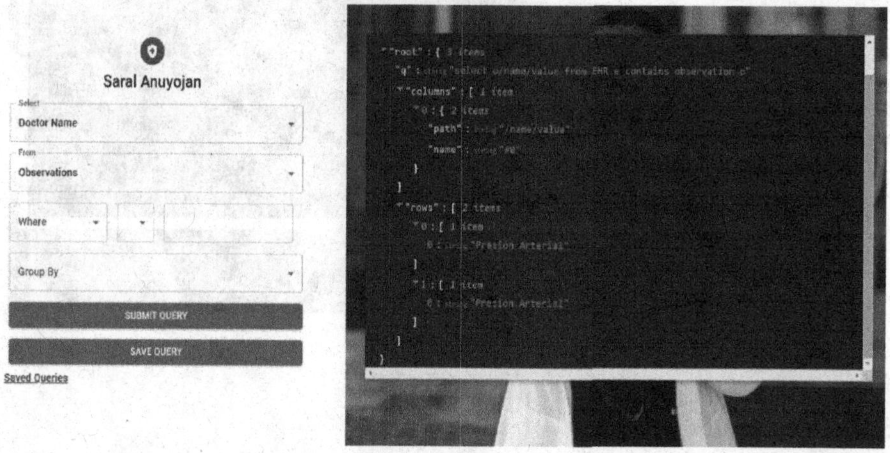

Fig. 7. Result for Q2

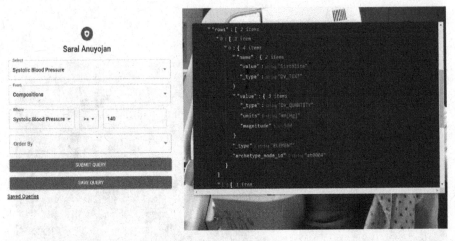

Fig. 8. Result for Q3

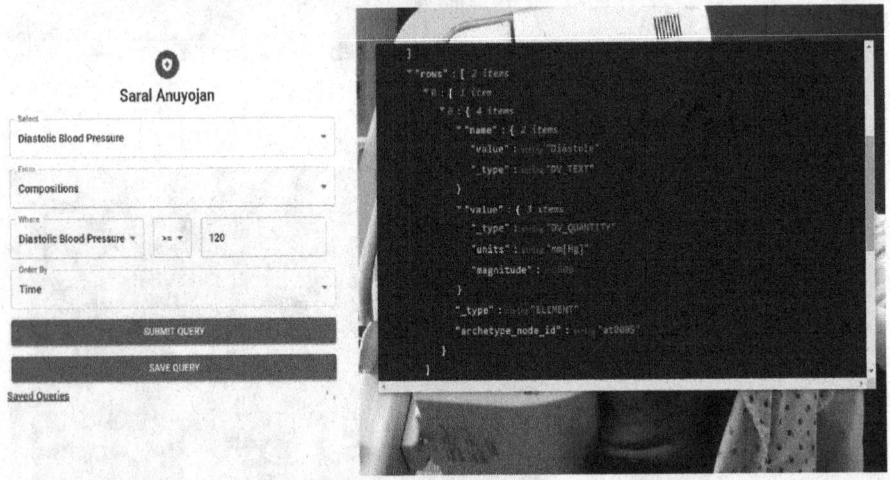

Fig. 9. Result for Q4

5 Conclusions

We proposed an easy-to-use query-based interface for non-technical users. The proposed interface performs almost all the required functions efficiently while being interactive and user-friendly. It is capable of translating the user input to AQL queries and returning the required output in a user-readable and comprehensive format.

References

1. Electronic Health Records I CMS. Electronic Health Records I CMS, 1 December 2021. www.cms.gov/Medicare/E-Health/EHealthRecords

2. MyHealthRecord I National Health Portal of India. MyHealthRecord I National Health Portal of India

3. Ministry of Electronics and Information Technology, Government of India. Ministry of Electronics and Information Technology, Government of India I Home Page. www.meity.gov.in

4. EHR Standards by Government of India. https://www.nhp.gov.in/NHPfiles/EHR-Standards-2016-MoHFW.pdf

5. Framework for Mobile Governance 2012. https://mgov.gov.in/Resources/Framework_for_Mobile_Governance.pdf

6. Digital India. www.digitalindia.gov.in

7. Beale, T., Heard, S.: openEHR. www.openehr.org

8. openEHR Architecture Overview: openEHR Architecture Overview. https://specifications.openehr.org/releases/BASE/Release-1.0.3/architecture_overview.html

9. Blobel, B., Pharow, P.: Archetypes and the HER. In: Advanced Health Telematics and Telemedicine: The Magdeburg Expert Summit Textbook, vol. 96, p. 238 (2003)

10. AQBE - QBE style queries for archetyped data - Scientific Figure on ResearchGate. https://www.researchgate.net/figure/Parameter-Blood-pressure-in-Archetype-form_fig1_258649971

11. Archetype Definition Language 1.4 (ADL1.4). Archetype Definition Language 1.4 (ADL1.4). specifications.openehr.org/releases/AM/latest/ADL1.4.html

12. Bray, T., et al.: Extensible markup language (XML) 1.0, 2000-10 (2000)

13. Iacob, I.E., Dekhtyar, A.: Towards a Query Language for Multihierarchical XML: Revisiting XPath. WebDB (2005)

14. Walmsley, P.: XQuery. O'Reilly Media Inc., Sebastopol (2007)

15. Alashqur, A.M., Su, S.Y.W., Lam, H.: OQL: a query language for manipulating object-oriented databases. In: Proceedings of the 15th International Conference on Very Large Data Bases (1989)

16. Sachdeva, S., Chu, W., Bhalla, S.: Relational-like query language for archetypes in standardized electronic health records databases. In: Proceedings of the ACM India Joint International Conference on Data Science and Management of Data (2019)

17. Kahng, M., et al.: Interactive browsing and navigation in relational databases. VLDB (2016)

18. Sachdeva, S., et al.: AQBE–QBE style queries for archetyped data. IEICE Trans. Inf. Syst. **95**(3), 861–871 (2012)

19. Sachdeva, S., Yaginuma, D., Chu, W., Bhalla, S.: Dynamic generation of archetype-based user interfaces for queries on electronic health record databases. In: Kikuchi, S., Madaan, A., Sachdeva, S., Bhalla, S. (eds.) DNIS 2011. LNCS, vol. 7108, pp. 109–125. Springer, Heidelberg (2011). https://doi.org/10.1007/978-3-642-25731-5_10

20. Sachdeva, S., Bhalla, S.: Implementing high-level query language interfaces for archetype-based electronic health records database. In: International Conference on Management of Data (COMAD) (2009)

21. Sachdeva, S., Batra, S., Bhalla, S.: Evolving large scale healthcare applications using open standards. Health Policy Technol. **6**(4), 410–425 (2017)

22. Reis, L.F., et al.: Integration through mapping—an OpenEHR based approach for research oriented integration of health information systems. In: 2018 13th Iberian Conference on Information Systems and Technologies (CISTI). IEEE (2018)

23. Ocean Health Systems, Sebastian Garde. Clinical Knowledge Manager. Clinical Knowledge Manager. https://ckm.openehr.org/ckm/

24. The Destination Point for Smart Approaches to Health Data. - Ocean Health Systems. Ocean Health Systems. https://oceanhealthsystems.com/

25. Archetype Designer. Archetype Designer. https://tools.openehr.org/designer

26. Kalra, D., Beale, T., Heard, S.: The openEHR foundation. Stud. Health Technol. Inform. **115**, 153–173 (2005)

27. EHRbase – Data Is for Life, Not Just for One System. EHRbase – Data Is for Life, Not Just for One System. https://ehrbase.org/

28. Vitasystems I Impressum I Vitagroup AG. Vitagroup AG. www.vitagroup.ag/de/ueber-uns/vit asystems

29. Peter L. Reichertz Institute for Medical Informatics: PLRI I Institute. Peter L. Reichertz Institute for Medical Informatics: PLRI I Institute. https://plri.de/en/institute

30. CaboLabs Health Informatics, Standards and Interoperability. CaboLabs Health Informatics, Standards and Interoperability. www.cabolabs.com

31. Gutierrez, P.P.: CaboLabs Health Informatics, Standards and Interoperability. CaboLabs Health Informatics, Standards and Interoperability. www.cabolabs.com/founder

32. Conrick, M.: Health informatics: Transforming healthcare with technology. CQUniversity (2006)

33. Mike Scaife, J., Rogers, Y.: Interactive representations. In: Cooperative Systems Design: A Challenge of the Mobility Age, vol. 74, p. 123 (2002)

34. https://ehrbase.org/

35. Pezoa, F., et al.: Foundations of JSON schema. In: Proceedings of the 25th International Conference on World Wide Web (2016)

36. Jacobson, D., Brail, G., Woods, D.: APIs: A Strategy Guide. O'Reilly Media Inc., Sebastopol (2012)

37. The API Design Platform and API Client. The API Design Platform and API Client – Insomnia. https://insomnia.rest/

Integrated Transmission and Distribution Modelling Using Multi-agent Based Framework

Devesh Shukla[1]([✉]) and S. P. Singh[2]

[1] School of Electrical and Electronics Engineering, Vellore Institute of Technology, Vellore, Tamil Nadu, India
devesh.shukla@vit.ac.in
[2] Department of Electrical Engineering, Indian Institute of Technology (BHU), Varanasi, India
spsingh.eee@iitbhu.ac.in

Abstract. Technological changes that enable the widespread adoption of renewable energy-based distributed generation and shift in the source of transportation sectors are introducing nascent challenges and opportunities in power generation, transmission, and distribution systems. Conventionally, the transmission and distribution systems were analyzed in a segregated manner under the assumption that the aggregate load subtended by the distribution systems at the transmission systems would remain constant, but with changing technologies and deployment of large-scale renewable sources of generation along with the increase in electric vehicle charging stations made the load subtended at the transmission level by distribution system to become uncertain with a probability of reverse power-flow. Therefore, an adequate method for analyzing the integrated transmission and distribution systems is to be evolved; in this article, we propose a multi-agent-based system framework for analyzing the integrated transmission and distribution system. The developed framework would be used to analyze the IEEE 9 bus test system at the transmission level and the IEEE 13 Bus test feeder at the distribution level.

Keywords: Integrated transmission and distribution ·
Multi-agent-based system · Active distribution system

1 Introduction

Conventionally the distribution system has been analyzed separately from the transmission systems, and different tools and methods were employed to investigate the transmission and distribution system [6,7]. The motive behind the segregated analysis of the Transmission System (TS) and Distribution System (DS) was derived from the observations that the aggregate load subtended by the DS at the TS generally happened to be balanced with a slow but predictable variation. This notion regarding the behavior of the DS began to be challenged owing to large-scale changes being implemented at the DS. These changes include changes like the change in the distribution system from a conventionally passive to an active distribution system, and transformation in the transportation sector from fossil fuel-based energy sources to electricity-based energy sources. A passive distribution system is one in which there is a prevalence of consumers (customers can only draw power from the grid). In contrast, in an active distribution system, consumers behave as prosumers injecting power back into the grid. The prosumers generally have localized power generation sources like rooftop PV to meet their needs and feed the excess power back to the grid. The usual DS that used to be unidirectional (i.e., power was fed from the transmission substation to the distribution substation) transformed into a more versatile system with probable backward flow of power from DS to TS. These changes mandate the necessity for having an adequate methodology for integrated analysis of transmission and distribution systems. The changing nature of the Transmission and Distribution $(T\&D)$ system is schematically inked in Fig. 1. The Fig. 1(a) illustrates the conventional scenario of TS and DS interaction, and Fig. 1(b) sketches the anticipated transformed behavior of TS and DS interaction in which power flow from DS to TS has also been inked. Methodologies and platforms that could be employed for co-simulation and $T\&D$ systems integrated studies for assessing the impacts of phenomenons happening at DS on TS and vice-versa are being developed. The investigation of economic aspects for the operation of $T\&D$ systems has been performed in [8] through a two-stage optimization-based method. The authors utilized the $T\&D$ economic models based on the residual demand curve and transmission-constrained supply curve. In [9], the global power flow method based on a master-slave splitting mechanism has been presented as a tool for ITD analysis. Authors in [4] propounded architecture for analyzing and monitoring the behavior of integrated $T\&D$ operation with a higher degree of DER penetration at DS levels. Another very interesting approach has been inked in [3], here existing simulators of TS and DS are linked using FCNS (Framework for Network Co-simulation), and the presented methodology could be utilized for analyzing the dynamic behavior of $T\&D$ system. Further, a scalable multi-time scale framework for $T\&D$ analysis using HELICS has been proposed in [1]. Although different tools and techniques are being developed and investigated for integrated analysis of TS and DS, the methodology for analyzing the integrated behavior of TS and DS systems for quasi-static analysis that may be utilized in applications such as ATC/ADC assessment are yet to be developed. In the present article, we would be detailing the formulation of a framework for integrated $T\&D$ studies that would employ a multi-agent-based mechanism.

2   Overview of Multi-agent Based System for Integrated TS and DS Analysis

A computerized system comprising multiple intelligent agents working in unison to solve a common problem is termed a Multi-Agent System (MAS). In our study, we would be promulgating a MAS-based framework for integrated analysis of TS and DS. We would be utilizing two different tools that are typically used for analyzing the transmission and distribution system separately. These tools would be interchanging the information at the 'PCC' (point of common coupling) through the multiple agents of MAS. MATPOWER [11] would be used

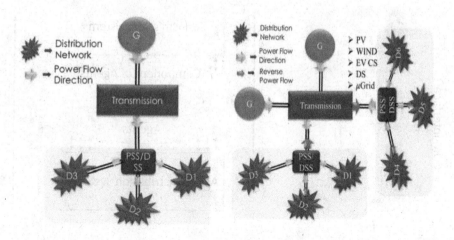

(a) Conventional transmission and distribution system.

(b) Modern transmission and distribution system

Fig. 1. Changing scenario of transmission and distribution systems.

Table 1. MAS implementation algorithm.

Step	Description
I	Model the TS system in TS solver
II	Identify the load nodes at which the DS are to be connected as ADS_{nodes} (PCC)
III	Assign agents for all the ADS_{nodes}/(PCC) of MAS
IV	Define the operating mode
V	Get the states i.e. (P, Q, V, δ, DS_{losses}, taps, caps)
VI	Establish the data exchange through MAS agent vectors using $agent2tr$ protocol

as a TS solver and OPENDSS [2] as DS solver. 'MATLAB' would be used to develop the MAS based framework. The MATPOWER is TS solver developed using 'MATLAB' language while the OPENDSS is an open-source tool developed specifically for distribution system analysis. In the proposed approach we would be utilizing a Component Object Model (COM) based interface for accessing the variables in OPENDSS. The schematic illustration of the proposed framework is shown in Fig. 2, the figure shows two parts, in (a) the solver's organization and COM interface have been inked while part (b) indicates the interaction between TS and DS through agent-based 'COM' interface. Two protocols namely ($agent2tr$/$tr2agent$/$dn2agent$/$agent2dn$ protocol) have been used for establishing interaction between TS and DS. Table 1 shows the implementation procedure for the proposed MAS-based mechanism.

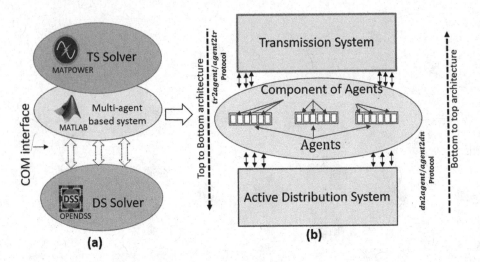

Fig. 2. Schematic illustration of the proposed framework.

The exchange of information between TS and DS is materialized through various agents α of the MAS-based system. The presented framework would work in two different configurations as given:-

- *Top-bottom approach*:- In this configuration, the solution to TS is obtained and it is communicated to the DS.
- *Bottom-up approach*:- Here the DS is solved first and the information is exchanged with TS for carrying out further analysis through the MAS agents.

The cardinality of the agents in the MAS-based system would be equal to the number of ADS_{node} in the system. Each agent would be a vector containing information and the status of different variables and components, for instance, α_i^j denotes the jth in the ith agent. Each agent would be having two major constituents namely TS agent vector and DS agent vector. state present in the DS connected to that ADS_{node}.

$$\alpha_i = \{\alpha_1, \alpha_2, \cdots, \alpha_a\}$$
$$\alpha_i = \{\alpha_i^1, \alpha_i^2, \cdots, \alpha_i^{a_i}\} \tag{1}$$

3 Modelling of the Active Distribution System and Integrated TS and DS System

The conventional DS is transforming into an Active Distribution System (ADS). In ADS, we have a mix of different sources (power producers) and sinks (power consumers) along with prosumers (consumers having to consume, produce and inject power into the grid). The ADS we will model in this work will comprise Solar PV-based distributed generation sources and prosumers. We will consider two types of prosumers; in the first category, it will be assumed that they have off-grid PV installation. In the second category, grid-connected systems with net metering provisions will be considered.

3.1   Solar PV-Based der Modelling

Modeling of the Solar PV has been done by considering it as a PQ load and designating P as $-ve$ for a generation. At the same time, Q would be $+ve$ or $-ve$ based on the operating scenario to which the smart invertor connecting the Solar PV to the grid would be subjected. The β distribution has been considered for modeling the *pdf* of solar irradiance on an hourly basis so as to incorporate the uncertainties associated with the PV power output [5]. The P and Q support that the smart inverter would provide by sourcing the PV module is given by

$$P^{PV} = P - P^{inv}_{losses} \qquad (2)$$

$$Q^{inv} \simeq Q \qquad (3)$$

$$P^{inv}_{losses} = (1 - \eta_{inv}) \cdot \sqrt{P^2 + Q^2} \qquad (4)$$

In the above equation, the η_{inv} is the operating efficiency of the inverter, and the Q^{inv}_{max} is the maximum reactive power support that the inverter could support.

3.2   Prosumer Modeling

In the Indian scenario, consumers with off-grid systems generally use them to meet their power requirements in case of unreliable power supply and lower their power consumption through the grid, reducing their energy costs. To model such behavior of the consumers, we have to consider their contracted load, load profile, and PV power profile.

$$P^i_{og}(t) = P^i_c - P^i_d(t) - Pg^i_{og}(t) \qquad (5)$$

The P_{og} would be subjected to the following constraints

$$0 \leq P^i_{og}(t) \leq P_c \qquad (6)$$

In the above equations, P^i_{og} is the power demand by the ith prosumers having P_c as its contracted load, P^i_d as its actual demand and Pg^i_{og} be the power produced by the off-grid system at time t. In DS the contracted load refers to the maximum rated demand (in kW, KVA, or HP) agreed to be supplied by the distribution company which the user can draw from the grid [10]. It is general practice that distribution companies penalize the users for violating the contracted load limit. Contrarily, for grid-connected systems installed through net metering schemes in some states of India (like UP), the prosumers could feed back to the grid. Such systems can be modeled using the following equations.

$$P^i_g(t) = P^i_c - P^t_d - Pg^i_g(t) \qquad (7)$$

The P_g^i would be subjected to the following constraints

$$- P_{pv}^i \leq Pg_g^i \leq P_{pv}^i \tag{8}$$

In the above equations, P_g^i is the power drawn by the ith grid-connected prosumer, Pg_g^i is the power generated by the PV installation, and P_{pv}^i is the installed capacity of the grid-connected system. The grid-connected systems with net metering or gross metering facilities are being offered in several states of India (like U.P) for loads starting from 1 kW to 2 MW. In such net metering provisions, the distribution companies bill the prosumers in terms of net energy injected into the grid. Such schemes are vital in promoting the large-scale deployment of grid-connected PV systems but in long run, they would limit the development of peer-to-peer energy trade through open access at distribution levels. Therefore, in long run, it can be expected that the new tariff policy would evolve as present tariff policies would not foster open access and competition in the distribution market.

3.3   TS and DS Modelling

The nodes in the TS to which the ADS would be connected are identified. The agents used in the MAS would equal the number of ADS nodes in the TS. The total load at the ADS node at TS level is decomposed into two parts, the first being the ADS node and the second fixed load. The mathematical modeling of the same is given as under

$$ADS_{load} = \zeta Pd^{DS} \tag{9}$$

$$Fixed_{load}^P CC = Pd_i^{PCC} - \zeta Pd^D S \tag{10}$$

$$\%ADS_{load} = \frac{\sum_{i=1}^{n_c} \zeta Pd_i^{DS}}{\sum_{i=1}^{n_c} Pd_i^P CC} \tag{11}$$

The ADS_{load} is the load subtended by the DS at the TS, ζ is number of ADS that form the DS which is connected to PCC, n_c refers to number of ADS nodes present in the system. The schematic illustration of the developed TS and DS integrated system is shown in Fig. 3.

4   Problem Formulation

The integrated TS and DS system has been solved by adhering to the power flow constraints of at both transmission and distribution levels. The power flow equations and constraints are given as under:

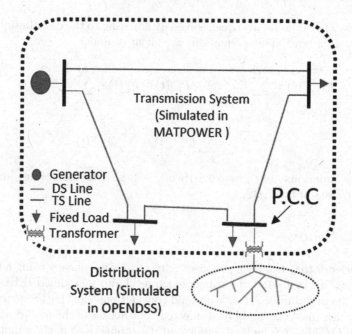

Fig. 3. Integrated TS and DS model.

4.1 TS Level

$$\sum_{i=1}^{Ng} Pg_i - \sum_{i=1}^{Nd} Pd_i - P_{loss} = 0 \tag{12}$$

$$\sum_{i=1}^{Ng} Qg_i - \sum_{i=1}^{Nd} Qd_i - Q_{loss} = 0 \tag{13}$$

$$P_i = \sum_{i=1}^{N} V_i V_j Y_{ij} cos(\theta_{ij} - \delta_i + \delta_j) \tag{14}$$

$$Q_i = \sum_{i=1}^{N} V_i V_j Y_{ij} sin(\theta_{ij} - \delta_i + \delta_j) \tag{15}$$

$$V_i^{min} \leq V_i \leq V_i^{max} \tag{16}$$

$$Pg_i^{min} \leq Pg_i \leq Pg_i^{max} \tag{17}$$

$$Qg_i^{min} \leq Qg_i \leq Qg_i^{max} \tag{18}$$

In the above equations, Ng is the number of generators, Nd is the number of load buses. N is the total number of buses in the TS system under study. Pg_i, Pd_i are active and reactive power generated by ith generator. V_i is voltage at ith bus, Y is the bus admittance matrix, θ_{ij} is the admittance angle of branch connecting bus i to j, and δ is the voltage angle.

DS Level. The ADS has also been solved by adhering to the distribution system power flow equations. These equations are given as under

$$Pd^{DS}(t) = \sum_{i=1}^{N_{pv}} P_i^{PV}(t) + P_i^{og}(t) + P_i^g(t) - \sum_{i=1}^{Nl} Pd_i^{DS}(t) \qquad (19)$$

$$Qd^{DS}(t) = \sum_{i=1}^{N_{pv}} Q_i^{PV}(t) - \sum_{i=1}^{Nd} Q_i^{DS}(t) \qquad (20)$$

The above equations are to be solved while adhering to the voltage and power flow constraints of the ADS.

5   Case Study

The framework that has been proposed is tested for its efficacy using a modified IEEE 9 bus test system at the transmission level and a modified IEEE 13-node feeder at the distribution level. We would consider bus 7 of the TS system as the PCC to which the ADS would be connected. The details of the modification done are shown in Table 2. The load profile and PV profile used in the simulation for the TS and ADS system is shown in Fig. 4a, and Fig. 4b. The load and PV profile used contains the hourly load variation at an interval of 15 min. The framework proposed in this article has been used to analyze the integrated operation of TS and DS. The total TS active power loss and voltage of the ADS node obtained during the integrated operation have been shown in Fig. 5a and b. The PV power injected by the off-grid rooftop PV and grid-connected rooftop PV has been inked in Fig. 6a and b respectively. From the plots, it can be seen that the power that is drawn/injected by the prosumers into the distribution grid from the roof-top PV systems depends on the type of the roof-top PV installation

Table 2. Description of the test system.

S.N	Modification done in TS				
	Bus No.	Fixed load		ADS load	
		P (MW)	Q (MW)	P (MW)	Q (MW)
1	7	82.67	24.49	17.33	10.51
–	Modification done in DS				
S.No	Grid connected rooftop-PV		Off-grid connected rooftop-PV		
	Bus No.	PV rating	Bus No.	PV rating	Connected load
1	675.1	485	652.1	120	128
2	675.2	68	670.1	15	17
3	675.3	290	670.2	60	66
4	611.3	170	670.3	100	117

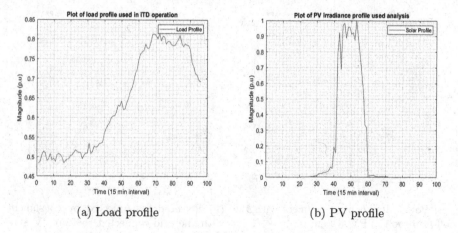

(a) Load profile   (b) PV profile

Fig. 4. The hourly load and PV profile for 24 h at 15 min intervals.

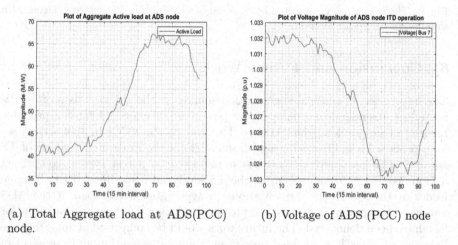

(a) Total Aggregate load at ADS(PCC)   (b) Voltage of ADS (PCC) node
node.

Fig. 5. The hourly active load and voltage magnitude of ADS (PCC) node during 24 h simulation at 15 min intervals.

they have. For instance, in Fig. 6a the minimum power exchanged with the grid is zero because the figure corresponds to off-grid PV system which does not have the provision of injecting power back to the grid. On the contrary, from Fig. 6b it can be seen that the total power drawn by the consumers becomes negative indicating that the power is being fed back from the prosumers to the grid.

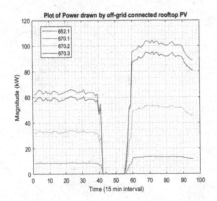 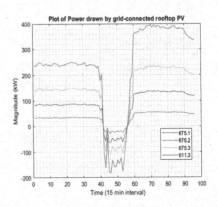

(a) Power drawn by consumers with the off-grid rooftop PV system.

(b) Power drawn/injected by consumers with the grid-connected rooftop PV system.

Fig. 6. The power drawn/injected by consumers with rooftop PV systems during 24 h simulations at 15 min intervals.

6   Conclusion and Future Work

The integration of renewable generation sources at the DS level as rooftop PV Systems (prosumers) and independent power producers could cause a probable reversal of power flow from ADS to TS through the PCC. In this article, we have proposed a multi-agent-based framework for the integrated analysis of TS and DS. The proposed method has been tested on a test system comprising a modified IEEE 9 bus system at the TS level and a modified IEEE 13 node feeder at the DS level. The variations in aggregated active load at the ADS node and the power drawn by the prosumers during the hourly simulation for 24 h have been delineated. The future work would be comprised of analyzing the behavior of blockchain-enabled peer-to-peer market operation and the effect of cyber-attack/contingency events on operational, control, and economic aspects of ITD operation.

References

1. Bharati, A.K., Ajjarapu, V.: A scalable multi-timescale T&D co-simulation framework using HELICS. In: 2021 IEEE Texas Power and Energy Conference (TPEC), pp. 1–6 (2021). https://doi.org/10.1109/TPEC51183.2021.9384985
2. Dugan, R.C.: OpenDSS Manual. Electric Power Research Institute (2016)
3. Huang, R., Fan, R., Daily, J., Fisher, A., Fuller, J.: Open-source framework for power system transmission and distribution dynamics co-simulation. IET Gener. Transm. Distrib. **11**, 3152–3162 (2017). https://doi.org/10.1049/iet-gtd.2016.1556
4. Jang, J.S.S.H.G.: Development of a transmission and distribution integrated monitoring and analysis system for high distributed generation penetration. Energies **10**(9), 1282 (2017)

5. Shukla, D., Singh, S.P., Mohanty, S.R.: Optimal strategy for ATC enhancement and assessment in presence of facts devices and renewable generation, pp. 1–6 (2018). https://doi.org/10.1109/NPSC.2018.8771774
6. Shukla, D.: ATC assessment and enhancement of integrated transmission and distribution system considering the impact of active distribution network. IET Renew. Power Gener. **14**, 1571–1583 (2020). https://digital-library.theiet.org/content/journals/10.1049/iet-rpg.2019.1219
7. Shukla, D., Singh, S.P.: Aggregated effect of active distribution system on available transfer capability using multi-agent system based ITD framework. IEEE Syst. J. **15**(1), 1401–1412 (2021). https://doi.org/10.1109/JSYST.2020.3000930
8. Singhal, N.G., Hedman, K.W.: Iterative transmission and distribution optimal power flow framework for enhanced utilisation of distributed resources. IET Gener. Transm. Distrib. **9**, 1089–1095 (2015). https://doi.org/10.1049/iet-gtd.2014.0998
9. Sun, H., Guo, Q., Zhang, B., Guo, Y., Li, Z., Wang, J.: Master-slave-splitting based distributed global power flow method for integrated transmission and distribution analysis. IEEE Trans. Smart Grid **6**, 1484–1492 (2015). https://doi.org/10.1109/TSG.2014.2336810
10. UPERC: UPERC (Rooftop Solar PV Grid Interactive Systems Gross/Net Metering) Regulations (2019). https://www.uperc.org/App_File/UPERCRSPV DraftRegulations2019-pdf116201865502PM.pdf
11. Zimmerman, R.D., Carlos, E.: User's Manual, MATPOWER. Vesion 7.1 edn (2020)

Incremental SVD-Based Hybrid Movie Recommendation to Improve Content Delivery Over CDN

Rohit Kumar Gupta[1], Yugam Shukla[1], Ankit Mundra[1(✉)], and Ritu Dewan[2]

[1] Manipal University Jaipur, Jaipur, India
`ankit.mundra@jaipur.manipal.edu`
[2] Galgotias College of Engineering and Technology, Greater Noida, India

Abstract. With the tremendous growth in the number of users watching on-demand movies over the Internet, Content Delivery Network (CDN) is used to provide a more efficient network and improve user experiences. CDN accelerates the response to end users after identifying the popularity of movie content. Nowadays, massive content has created difficulties in accurately predicting each movie's popularity. A recommendation system uses filtering tools that recommend content according to the popularity among the users. Currently, recommendation systems play a vital role in identifying the relevant content from massive data sources and providing options to users to select content according to interest. To deal with related content, we analyze different algorithms utilized in popular recommendation systems and develop a hybrid system. We are using a hybrid approach by merging basic content based with Incremental SVD-based filtering to enhance the effect of the recommendation system for movie content to the end users over CDN.

Keywords: Recommendation systems · Content filtering · Collaborative filtering · SVD · Incremental SVD · Content delivery network

1 Introduction

In recent years, e-commerce firms like Amazon, Netflix, and YouTube are utilizing Content Delivery network platforms to provide better end-user experiences in watching on-demand movies and video content. Recommendation systems have been a hot research topic due to their vast application in video and OTT platforms. They are utilizing recommendation algorithms [2] to assist customers in finding content that interests them. These contents are pre-fetched by the CDN Proxy servers because of their significant relationships between content metadata and user interest. Now a days, These relations have become essential factors in content allocation in content provision sites. Based on these relations, the recommendation system generates appropriate suggestions.

In the movie recommendation system, distinct types of movies are recommended to the end users because of distinctive features such as movies watched, comments, liked, disliked, and rated by the user, which may be used to generate more precise suggestions.

S. Sachdeva et al. (Eds.): BDA 2022, LNCS 13830, pp. 188–195, 2023.
https://doi.org/10.1007/978-3-031-28350-5_15

A large number of movies and user profile features created problems in identifying the best recommendations to organize and cache the movies at proxy servers. To overcome these problems, movie recommender systems are used. Users might be suggested a collection of movies depending on their interests or the notoriety of the films.

Different Filtering techniques are applied to develop recommendation systems. Content-based filtering techniques [6] compare the item's content and a user profile to recommend things. But collaborative filtering [5, 6] method for generating automatic suggestions based on user's activities and interests by combining the preferences data of more users. Meanwhile, it is based on past performance instead of just context. Different Memory-based and model-based collaborative filtering approaches are used to perform the recommendation.

There needs to be more than the single approach mentioned above in recommendation systems to provide efficient results. In our model, we present a computing technique that maintains the simplicity of previous methods while requiring minor parameter modification, resulting in a slight increase in total complexity determined by the input data structure. We compare the various algorithms utilized in popular recommendation systems and develop a hybrid system.

CDN system prefers recommendations related to the content available in the cache. These recommendations may utilize users' interests to generate content provider (CP) revenues. We analyze and compare the content-based and collaborative filtering approaches such as Item-Item, User-Item, SVD, and SVD++ with the hybrid system on Movielens dataset [19]. The hybrid model was constructed by combining content-based and SVD-based filtering to enhance the accuracy of movies recommended to the user. In this work, we are analyzing different recommendation techniques with a hybrid approach and propose designing a hybrid recommendation policy that is mutually beneficial. To prefetching of content on proxy servers of CDN is based on a recommendation to end users according to the proposed hybrid model.

2   Background and Related Work

2.1   Content Delivery Network

CDN uses distributed infrastructure in proxy servers at various locations for content delivery and replicating content in proxy servers, as shown in Fig. 1. Recently, identifying related content based on user interest and replicating it at appropriate locations are primary tasks to achieve better Quality of Service (QoS). Recommendation technologies [24] help reduce the content delivery cost for reaching its end users. Recommendations prefetched in cached items [27] can lead to increased cache hits and reduced delivery costs.

2.2   Recommendation Techniques

Content-based filtering techniques [6] compare the item's content and a user profile to recommend things. A collection of descriptors or terms denotes each content of an item, and these are often words found in a text. To compare the relative value of different films, Term Frequency (TF) and Inverse Document Frequency (IDF) concepts are used in content-based filtering techniques. You will come across users who have not rated a substantial number of goods in many recommender systems [20]. Methods that rely substantially on data, such as KNN [23] and SVD, perform badly for these users. Item content information is utilized to recommend items to consumers.

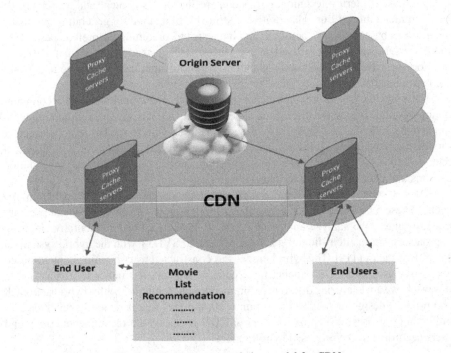

Fig. 1. Hybrid recommendation model for CDN

Collaborative filtering [5, 6] is a technique for automatically predicting a user's interests by combining preference data from various users. It is based on the concept that if two users have related views on an issue, X is likelier to share B's viewpoint on a different problem than a arbitrarily picked individual. In the context of a recommendation system [10], movie preferences might offer predictions about which movie a user should enjoy based on the likes of other movies, given a partial list of that user's preferences based on features such as like, dislike, etc.

Matrix factorization (MF), which is entirely based on dimensionality reduction and latent variable decomposition, is used in Model-based Collaborative Filtering [17]. MF is commonly used in recommendation systems because it can deal with scalability and sparsity better than other algorithms. MF aims to learn the user's latent preferences and the item's latent characteristics from existing ratings and then predict unknown ratings using the dot product of the user's and object's latent characteristics.

The Singular value decomposition (SVD) [14] is a collaborative filtering approach used to design a recommender system. The matrix is designed with each column indicating an object and each row corresponding to a user. The users give a rating to items are the values of this matrix. SVD is a technique for breaking down a matrix into three smaller matrices. This matrix factorization approach used in the SVD approach decreases the number of features in a dataset by lowering the space dimension from N to a smaller integer K. To estimate the user-item rating matrix [18], they used models inspired by SVD. SVD models are widely popular due to their accuracy and scalability.

Alternatively, item-item filtering resolves this problem by identifying relationships between things and recommending appropriate goods to consumers. Moreover, SVD and SVD++ use a dimensionality reduction approach with matrix factorization to the problem [13, 14]. The usage of implicit feedback significantly influenced the findings as essential as obtaining the Boolean vector of whether a user rated a video. When more concrete implicit feedback models, such as the customer's purchase history, become accessible, this might lead to even more exciting results. These approaches are adept at predicting ratings for movies not seen yet by users.

Amazon was the first to invent and use Item-Item Collaborative [7] Filtering in 1998. Item-to-item collaborative filtering connects each of the user's purchased and rated things to similar things, then aggregates those similar things into a suggestion list rather than matching the user to comparable consumers. Their similarity is determined according to the distance between things. For a movie recommendation system, the user's rating might be used to define the vector for each movie, and these vectors are compared later using distance measures using different approaches like Jaccard Similarity, Cosine distance, Euclidean distance, or Pearson correlation.

2.3 Hybrid Approach for Recommendation

There needs to be more than the single approach mentioned above in recommendation systems to provide efficient results. The different hybrid approaches are proposed in the literature. In [3, 25], Hybrid SVD is presented, which shows more information is required other than collaborative information to recommend the movie items. In [8], an integrated framework combines rating prediction and domain detection models to design a hybrid scalable collaborative filtering-based recommendation system. [20] calculates the similarity between items and users. They consider metadata of the movies to generate information for rating. The method proposed in [21] combines multiple ratings estimated by various similarity measures and offers high-quality personalized suggestions to the target user. The algorithm proposed in [22] uses movie recommendations to generate a recommended list through singular value decomposition, collaborative filtering, and cosine similarity. The proposed method MR-TWOA-based recommendation system [28], is validated in terms of mean absolute error, precision, and recall on the publicly available movie-lens dataset.

3   Methodology for Recommendation System

A hybrid model that combines two separate models can address the limitations of a single model. The SVD model predicts the user's rating for the suggested movie based on the movie recommended [26] by Content-Based Filtering. We arrange the movies in

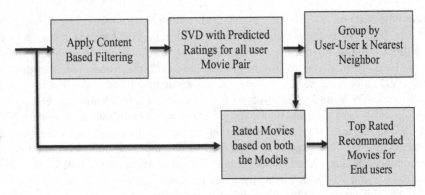

Fig. 2. Hybrid recommendation model

decreasing order of SVD + + projected ratings, then group them by User-User k nearest neighbor [23]. The hybrid model generates more precise and appropriate results.

When more concrete implicit feedback models, such as the user's purchase history, become accessible, this might lead to even more exciting results. As shown in Fig. 2, the proposed hybrid model filters the content again and provides reliable recommendations to end users. Although this adjustment appears to be small in scale, it significantly influences the top ten movie suggestions for a targeted user. In the process of cache allocation and content placement in the proxy servers of the CDN, the proposed scheme allows the CDN to decide the recommendations from the cached contents. The different users received a recommendation list and accessed content as per the list, which decreased the delivery cost.

4   Implementation and Results

In this paper, we used the MovieLens 100K dataset to investigate the issue and evaluate the results. The information was gathered via the MovieLens source [19]. We applied the proposed model in the data set and compared the result with the basic models Content-Based User-User Collaborative and Item-Item Collaborative. RMSE [26] is used as evaluative criteria in the results.

$$RMSE = \sqrt{\frac{1}{k} \sum_{u.i} (PR_{u.i} - AR_{u.i})^2} \qquad (1)$$

where, $PR_{u.i}$ indicates predicted rating of user u to the movie i, and $AR_{u.i}$ indicates actual rating of user u to the movie i.

The above Table 1 and Fig. 3 are showing the results obtained for different methods using the RMSE value (Root Mean Squared Value). We can also state that the RMSE value needs to be lesser for a model to be better than the other models. According to the above table, our Hybrid Model is better than all others due to lower RMSE values.

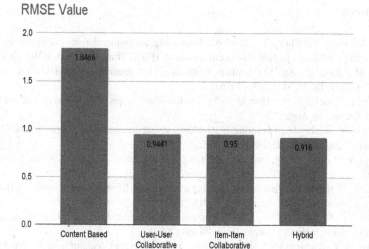

Fig. 3. RMSE Value for different Approaches on dataset

Table 1. Results on the Dataset

Method	RMSE Value
Content-Based	1.8466
User-User Collaborative	0.9441
Item-Item Collaborative	0.9500
Hybrid	0.9160

5 Conclusion

Now a days, Recommendation systems are part of different applications such as OTT, eCommerce websites, etc. Showing the proper suggestions improves user engagement, which has the advantage of increasing income. According to the proposed hybrid method in this paper, one system can include a mixture of methods related on various circumstances. As a result, if we have metadata on movies, such as genre, artists, and production firms, we may recommend movies to customers based on that information. The proposed Hybrid model filters the content and provides reliable recommendations to end users. Although this adjustment appears to be small in scale, it significantly influences the top ten movie suggestions for a targeted user. CDN Caching allocation and cache deployment based on recommendations have an improvement on the resulting delivery cost. Therefore, a direction for future work would be to model a CDN for caching and guidance to achieve better Quality of Service (QoS).

References

1. Tan, X., Guo, Y., Chen, Y., Zhu, W.: Improving recommendation via inference of user popularity preference in sparse data environment. IEICE Trans. Inf. Syst. **E101–D**(4) (2018)
2. Yan, M., Sang, J., Xu, C.: Unified YouTube video recommendation via cross-network collaboration. In: ACM ICMR (2015)
3. Frolov, E., Oseledets, I.: Hybrid SVD: when collaborative information is not enough. In: ACM Recsys, 13 August 2019
4. Deldjoo, Y., Elahi, M., Cremonesi, P.: Using visual features and latent factors for movie recommendation. In: CBRecSys, 16 September 2016
5. Koren, Y.: Collaborative filtering with temporal dynamics. Commun. ACM **53**(4), 89–97 (2010)
6. Ricci, F., Rokach, L., Shapira, B.: Introduction to the Recommender Systems Handbook. Springer, Boston (2011). https://doi.org/10.1007/978-0-387-85820-3_1
7. Linden, G., Smith, B., York, J.: Amazon. com recommendations: item-to-item collaborative filtering. IEEE Internet Comput. **7**(1), 76–80 (2003)
8. Gunjal, S.N., Yadav and, S.K., Kshirsagar, D.B.: A hybrid scalable collaborative filtering based recommendation system using ontology and incremental SVD algorithm. In: 2020 International Conference on Smart Innovations in Design, Environment, Management, Planning and Computing (ICSIDEMPC), pp. 39–45 (2020). https://doi.org/10.1109/ICSIDEMPC 49020.2020.9299604
9. Tsolakidis, A., Triperina, E., Sgouropoulou, C., Christidis, N.: Research publication recommendation system based on a hybrid approach. In: Proceedings of the 20th Pan-Hellenic Conference on Informatics. Association for Computing Machinery, New York, NY, USA, Article 78, pp. 1–6 (2016)
10. Hande, R., Gutti, A., Shah, K., Gandhi, J., Kamtikar, V.: Moviemender - a movie recommender system. Int. J. Eng. Sci. Res. Technol. (IJESRT) **5**, 11 (2016)
11. Ma, H., Zhou, D., Liu, C., Lyu, M.R., King, I.: Recommender systems with social regularization. In: ACM International Conference, February 2011
12. Mohd Kasirun, Z., Kumar, S., Shamshirband, S.: An effective recommender algorithm for cold-start problem in academic social networks. Math. Probl. Eng. **2014**,11 (2014)
13. Katz, G., Shani, G., Shapira, B., Rokach, L.: Using Wikipedia to Boost SVD Recommender System, 5 December 2012. 19:03:39 UTC
14. Yan, M., Shang, W., Li, Z.: Application of SVD technology in video recommendation system. In: IEEE, June 2016
15. Osmanli, O.N.: A singular value decomposition approach for recommendation system, July 2010
16. Sarwar, B., Karypis, G., Konstan, J., Riedl, J.: Application of dimensionality reduction in recommender system -- a case study. In: WebKDD-2000 Workshop (2000)
17. Lemire, D., Maclachlan, A.: Slope One Predictors for Online Rating-Based Collaborative Filtering, 9 January 2012
18. Gupta, R.K., Verma, V.K., Mundra, A., Kapoor, R., Mishra, S.: Improving recommendation for video content using hyperparameter tuning in sparse data environment. In: Nanda, P., Verma, V.K., Srivastava, S., Gupta, R.K., Mazumdar, A.P. (eds.) Data Engineering for Smart Systems. LNNS, vol. 238, pp. 401–409. Springer, Singapore (2022). https://doi.org/10.1007/ 978-981-16-2641-8_38
19. Movielens Dataset, December 2019. https://grouplens.org/datasets/movielens
20. Ricci, F., Rokach, L., Shapira, B.: Recommender systems: introduction and challenges. In: Recommender Systems Handbook, pp. 1–34 (2015)

21. Pirasteh, P., Bouguelia, M.-R., Santosh, K.C.: Personalized recommendation: an enhanced hybrid collaborative filtering. Adv. Comput. Intell. **1**(4), 1–8 (2021). https://doi.org/10.1007/s43674-021-00001-z
22. Bhalse, N., Thakur, R.: Algorithm for movie recommendation system using collaborative filtering. In: Proceedings of Materials Today (2021)
23. Zhang, H.-R., Min, F., He, X., Xu, Y.-Y.: A hybrid recommender system based on user-recommender interaction. Math. Probl. Eng. **2015**, 1–11 (2015). https://doi.org/10.1155/2015/145636
24. Kaafar, M.A., Berkovsky, S., Donnet, B.: On the potential of recommendation technologies for efficient content delivery networks. ACM SIGCOMM Comput. Commun. Rev. **43**(3), 74–77 (2013). https://doi.org/10.1145/2500098.2500109
25. Salmani, S., Kulkarni, S.: Hybrid movie recommendation system using machine learning. In: International Conference on Communication information and Computing Technology (ICCICT) 2021, pp. 1–10 (2021). https://doi.org/10.1109/ICCICT50803.2021.9510058
26. VahidiFarashah, M., Etebarian, A., Azmi, R., EbrahimzadehDastjerdi, R.: A hybrid recommender system based-on link prediction for movie baskets analysis. J. Big Data **8**(1), 1–24 (2021). https://doi.org/10.1186/s40537-021-00422-0
27. Tsigkari, D., Iosifidis, G., Spyropoulos, T.: Split the cash from cache-friendly recommendations. In: 2021 IEEE Global Communications Conference (GLOBECOM), pp. 1–6 (2021).https://doi.org/10.1109/GLOBECOM46510.2021.9685088
28. Tripathi, A., Mittal, H., Saxena, P., Gupta, S.: A new recommendation system using map-reduce-based tournament empowered Whale optimization algorithm. Complex Intell. Syst. **7**(1), 297–309 (2020). https://doi.org/10.1007/s40747-020-00200-0

Business Analytics

Improving Emotional Confusions in SNS Sentiment Analysis by Partial Redistribution of BERT Discrimination Results

Yenjou Wang[1] and Qun Jin[2(✉)]

[1] Graduate School of Human Sciences, Waseda University, Tokorozawa, Japan
yjwjennifer2021@ruri.waseda.jp
[2] Faculty of Human Sciences, Waseda University, Tokorozawa, Japan
jin@waseda.jp

Abstract. Most of previous research works of sentiment analysis on SNS mainly focused on polarity analysis to probe into user tendencies. However, human emotions are complex and changeable. It is difficult to use the results of traditional polarity analysis in the real-time application services. Although finer-grained sentiment analysis may provide more detailed results, it has the problem of ambiguity in the definition of features between emotions. In this study, we propose a partial redistribution method based on BERT to tackle this problem. It improves emotional confusion in sentiment analysis through the confusion matrix, and further uses binary classification models to re-train the data of the confused emotional group through the redistribution process. In addition, the model makes it possible to re-extract and define features for specific emotions. Finally, F1 score is used to judge whether each feature correction process exerts positive impact on the model. Experimental results demonstrate that our proposed approach is effective in improving emotional confusion issues in SNS sentiment analysis.

Keywords: Sentiment analysis · Finer-grained analysis · Emotion model · BERT · Machine learning

1 Introduction

Over the past few years, various mobile devices, such as mobile phones and wearable devices, have been developed rapidly, and social networking services (SNS) users and their posts also have increased greatly. The data on SNS is extensively used for maintaining relationship with friends. Meanwhile, the emotions in the posts also express the cognitive tendencies in various fields, such as economy, society, and culture. Consequently, SNS analysis has become an important research domain in sentiment analysis (SA) over the past decades.

SA is a common task in the field of Natural Language Processing (NLP), which aims to detect emotions in text. The classical document-level sentiment analysis focuses on determining the user's emotional polarity (e.g., positive, negative, and neutral) towards a product or event from the whole sentence [1, 2], and using feature engineering to optimize the performance of classifiers (such as support vector machines) [3]. However, people's

S. Sachdeva et al. (Eds.): BDA 2022, LNCS 13830, pp. 199–210, 2023.
https://doi.org/10.1007/978-3-031-28350-5_16

emotions are complex, including happiness, sadness, love, anger, fear, surprise, and trust, etc. It is highly possible for a text to contain multiple emotions. For the behavior-specific real-time services, such as chatting robots, and medical monitoring software (for infants or psychiatric patients), tedious and time-consuming feature engineering is not enough to satisfy their needs.

Previous research on the finer-grained SA, such as multi-label or multi-category SA, mainly uses neural network models, such as RNN [4], CNN [5] and LSTM [6]. Neural networks can learn representations from data without complex feature engineering. Particularly, the deep learning has become the most popular research method in this field. Nevertheless, since the data on SNS contains multiple emotions and noise, many studies have been committed to increasing the feature extraction rate of the model based on neural networks, such as with the aid of emojis or the combination of multiple neural networks [7, 8]. However, for SA, finer-grained classification means that a more detailed definition is needed for each emotion, and feature discrimination is also more difficult in the model training. It leads to precision rate of existing studies mostly in the range of 50–70%, which is one of the primary problems in the current SA.

This paper proposes a partial redistribution method, which employs Bidirectional Encoder Representations from Transformers (BERT), a sentence-level pre-training model [9], as the basis for a SA model. BERT performs masked word prediction on a large amount of untagged text through a masked language model, which focuses on the context, and enables the model to obtain a generic emotion feature value. Then the confusion maps are used to analyze the confusion among emotions in the BERT model, and a partial redistribution method is used to resolve the confused emotion issue. Moreover, several binary classification models are applied to re-train the confused emotion group so as to enhance the ability of identifying the features by the model.

The remainder of this paper is organized as follows. In Sect. 2, the related work on the finer-grained SA on SNS and the SA by BERT are overviewed. In Sect. 3, the process and method of research architecture are described in detail. In Sect. 4, we present and evaluate the experiment results of our proposed model and discuss the results. Finally, the conclusion is given, and future work directions are highlighted in Sect. 5.

2 Related Work

The traditional SNS sentiment analysis studies mostly explore user tendencies through polarity analysis, but the complexity of human language makes it difficult to express meaning only through polarity. For some real-time applications, such as chat software and body status feedback software, the results of polarity analysis are difficult to meet their needs. Therefore, research works have been tried to extensively explore the field of finer-grained SA. I. Tripto et al. [10] used CNN and LSTM respectively to detect multi-categorical emotions on YouTube. In addition, Z. Jin et al. [11] used BERT for analysis and proposed an improved TF-IDF method to show the importance of words in multi-label classification problems through the calculation of different full scales. H. Fei et al. [12] proposed a topic-enhanced capsule network, which could obtain potential topic information and capture corresponding emotional features for multi-label emotion detection tasks. However, improving correctness is still a major difficulty for multi-categorical and multi-label.

In SA, the semantic problem frequently arises due to different users having different perceptions of the same vocabulary. With the release of BERT, this problem has been solved. BERT is pre-trained from a large corpus. Even if the same word exists in different paragraphs, BERT can know whether these words have the same meaning. Therefore, many researchers use BERT and extension methods for more detailed sentiment analysis tasks. T. Tang et al. [13] fine-tuned BERT. Thus, their BERT model can be applied to multi-label sentiment analysis to deal with complex emotional issues in SNS content due to language translation. In addition, X. Li et al. [14] used a method called GBCN, which includes context-aware aspect embeddings to enhance and control the representation of BERT. Through principal component analysis (PCA), Song et al. [15] visualized and explored the potential of BERT intermediate layers and enhanced BERT fine-tuning performance. Despite this, few literatures have explored the confusing status of BERT for categories. As the confusion of categories may be related to the precision rate of the model, this study aims to propose a method to solve the confusion.

3 Framework of Partial Redistribution for Improving Emotional Confusions

This section presents and discusses the methods used in our model, which aims to solve the confusion problem in fine-grained sentiment analysis in social networking services. We first describe the use of BERT for general SA tasks, introduce the method for calculating the mixed emotions model, and explain how our proposed partial redistribution method can enhance the ability of the BERT model to identify the characteristics of mixed emotions. Our approach consists of four main parts: data preparation, BERT-based SA model architecture, emotional confusion confirmation, and partial redistribution and calibration. An overview of the proposed framework is shown in Fig. 1.

Section 3.1 describes which BERT architecture is used in this study and the process of fine-tuning to meet the analysis goals. Section 3.2 describes the calculation method and process by confusion matrix to pick up the confused emotion group in SA. Section 3.3 describes the process of the partial redistribution method. The BERT model redefines and adds the feature discrimination ability of the confused emotion group through the partial redistribution method.

3.1 Model Architecture for Sentiment Analysis Based on BERT

Since SNS posts are unstructured data, they often contain purely semantic issues in different contexts. SNS content is characterized by its large quantity and rapid growth. Therefore, traditional feature engineering can be time-consuming. To address this issue, we modify the bert-base-uncased model for SA. Through BERT's training tasks of masking word prediction and next sentence prediction, BERT can consider the context and generate general features for future use. In contrast to conventional two or three dimensional polarity analyses, the output of our task is an 8-dimensional vector. Therefore, based on the pre-trained model, we add an 8-dimensional linear classifier for fine-tuning and use the sigmoid cross-entropy loss function to support multi-label classification. At the same time, we use AdamW [16], which is designed to improve the generalization

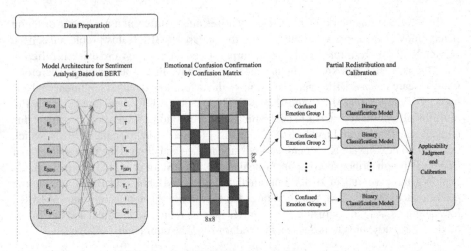

Fig. 1. Overview of the proposed framework

ability of the model through weight decay regularization that helps to prevent overfitting, and a learning rate scheduler as the optimizer to set the learning rate.

3.2 Emotional Confusion Confirmation

If emotions are misjudged or confused during the sentiment analysis process, it means that the definition of several emotional features in the model is unclear or similar. If the confusion can be improved, the discrimination ability of the model will be improved. In this section, we focus on describing how to identify the confusion parts from the analysis results. Section 3.3 will describe the methods to improve the confusion. The confusion matrix is a visualization tool. Each column of the matrix represents the model prediction of a category. Each row represents an instance of a category. Given the intersection of levels, it can be seen whether the model confused two different categories, e.g., data of a cat is confused with a dog in the model's prediction. In addition, all the correct predictions are located on the diagonal of the table, while the incorrect predictions are expressed as values outside the diagonal. The following are the four basic elements that make up the confusion matrix. All basic elements can be calculated as the model metrics, such as accuracy, precision, recall, F1 score, etc. The definition and examples of confusion matrix elements are shown in Table 1.

- TP (True Positive): Positive samples are predicted to positive correctly.
- TN (True Negative): Negative samples are predicted to negative correctly.
- FP (False Positive): Negative samples are mis-predicted as positive samples.
- FN (False Negative): Positive sample are mis-predicted as negative samples.

In traditional machine learning, TP and TN rates are often the focus, as a high TP or TN value indicates that the model has a high accuracy. However, in this study, we will focus on false positive (FP) and false negative (FN) rates. The confusion matrix for

Table 1. Definition and examples of confusion matrix.

Confusion Matrix		Actual Values	
		Positive Cat	Negative Dog
Predicted	Positive Cat	TP (True Positive) 5	FP (False Positive) 3
	Negative Dog	FN (False Negative) 2	TN (True Negative) 3

multi-class classification is used to determine the error rate for each emotion group. The formula for the error rate can be represented as Eq. (1).

$$ErrorRate = 1 - \text{Accuracy} \tag{1}$$

where the accuracy can be represented as Eq. (2).

$$\text{Accuracy} = \frac{TP + TN}{TP + TN + FP + FN} \tag{2}$$

3.3 Partial Redistribution and Calibration

As mentioned previously, if the model struggles to classify emotions accurately, it may be because the emotional features have not been properly learned and classified. Our first goal is to give confused emotions a chance to be re-learned and have their characteristics redefined. We propose a partial redistribution method that considers the training cost for the confounded emotion combinations during the re-learning phase. Since the confusion calculation is based on 8-dimensional BERT results in Sect. 3.1. If this analysis is used as input for the relearning phase, there are two problems.

(1) Input all the analysis results: wrong discrimination may lead to a bullwhip effect to cause more bias in training, and
(2) Input only the correct analysis results: this will reduce the data available for training.

To address these issues, we redistribute the data of all confusing sentiment groups in this stage. In finer-grained sentiment analysis training, an emotion that the model clearly discriminates increases the probability of other emotions being correctly classified. Therefore, to maximize the model's overall performance, we hypothesize that sentiment groups with a higher degree of confusion (higher error rate) should be reallocated first. For this reason, the results from Sect. 3.2 are sorted in descending order and partially re-extracted from the dataset. These results are then redistributed into binary classification models to re-train and extract new features. This method allows us to reassign a portion of the data that was incorrectly judged due to confusion in the first stage to the correct emotion category. The process of the partial redistribution method is shown in Fig. 2.

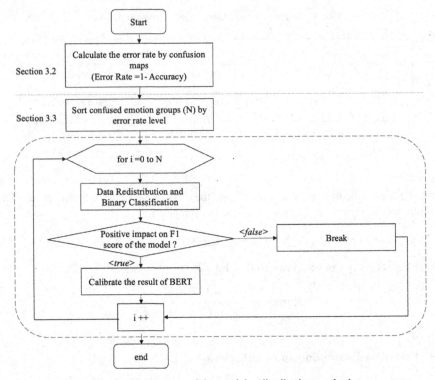

Fig. 2. The process of the partial redistribution method

Moreover, we consider that not all calibrations will have a positive effect on the original BERT model. Therefore, we use the F1 score, a model accuracy metric that takes into account both precision and recall, as a measure of the positive impact of binary classification in this study. Each binary classification result will be used to calibrate the original BERT model and determine whether the F1 score has a positive impact on the model's performance. By doing this, we can ensure that the model's discriminative ability is improved.

4 Experiment Results and Discussion

4.1 Experiment Setup

This is a highly integrated development model based on Python's Juniper Notebook, which was built based on Python 3.7, NumPy 1.19.5, scikit-learn 1.0.2, transformers 4.16.2, and a randomly assigned GPU (Tesla P100 or RTX6000) for training.

In the training process, two AdamW and two learning rate schedules are used to be optimizers one set for BERT model and another for the linear classifier of the output layer. Finally, we use seed = 20, and epochs = 10 to start our BERT fine-tune flow, and considering the overfitting problem, we set a dropout rate of 0.1.

4.2 Data Preparation

We collected data from the dataset proposed in [17]. GoEmotions is a human-annotated dataset of 58k Reddit comments extracted from English-language subreddits and labeled with 28 emotion categories, including 12 positives, 11 negatives, 4 ambiguous emotion categories, and 1 neutral. To maintain the emotional extensibility and balance in our model, the Plutchik's wheel of emotion [18] is used as the basis of emotion theory in this study. There are nine emotions that can be extracted from the GoEmotions dataset and mapped to the Plutchik's wheel of emotion. Since the same type of emotion on the wheel of emotions contains various emotions of different emotional intensities, emotion categories cannot be directed mapping, such as anticipation, which is composed of curiosity and caring. The noise data, e.g., gibberish, emoji, and a post containing only the mask of "[NAME]" and punctuation, is removed. During the extraction process, in addition to manually deleting garbled characters and blank data, we further use fuzzy analysis by Microsoft Excel to remove over-similarity and under-featured sentences for pre-processing. Finally, we define a dataset containing eight emotion categories (joy, surprise, anger, sadness, fear, trust, and anticipation), which contains 22,639 posts in total. The corpus is split into training, test, and validation datasets in the ratio 80:10:10. In each training, maintaining both the proportion of the classes is by random sampling.

4.3 Results and Discussion

In addition to the four fundamental evaluation indicators commonly used in machine learning (training accuracy, training loss, validation loss, and validation accuracy), precision, recall, and the F1 score are often chosen as evaluation indicators for sentiment analysis problems. In this experiment, the F1 score is used as a critical evaluation index because it takes both precision and recall into account.

Original BERT Implementation Result. As the current mainstream NLP model, BERT has strong semantic analysis capabilities by considering context and pre-training methods and has achieved good results in many NLP tasks. However, multi-category SA is also a complex problem for BERT. This study randomly selected 80% of the posts of the dataset for training, 10% for testing, and 10% for validation. Tables 2 and 3 summarize the performance of our best original model, BERT, on the validation set, which achieves the training accuracy rate of 78.12%; the validation accuracy rate reaches 63.62%, and the average F1 score of validation achieves 0.63. The learning curves are shown in Fig. 3. In addition, the solid and dashed curves in the left figure show the training and validation accuracies. The solid and dashed curves in the right picture show the training and the validation loss.

Table 2. The results of BERT.

	Accuracy	Loss
Training dataset	78.12	1.49
Validation dataset	63.82	1.63

Table 3. F1 score results table of the overall model and each emotion.

	F1 score for test data	F1 score for validation data
Anger	**0.79**	0.61
Fear	0.73	0.55
Anticipation	**0.83**	**0.77**
Surprise	0.71	0.56
Joy	**0.84**	**0.71**
Sadness	**0.80**	0.61
Trust	0.79	0.60
Disgust	0.61	0.42
Overall	**0.76**	**0.63**

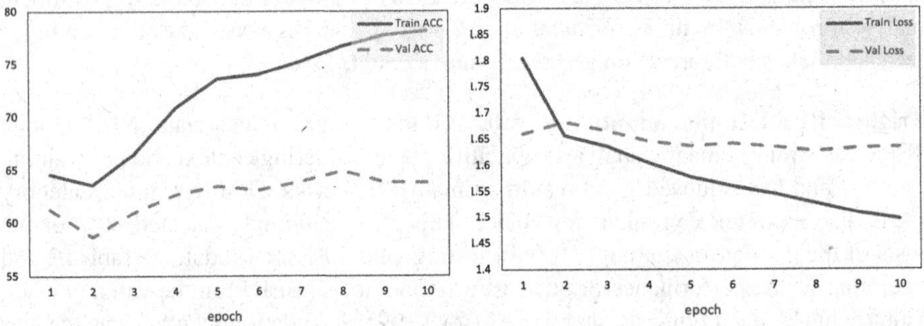

Fig. 3. Training Curve Graph of Accuracy and Loss

The result of the confusion matrix is shown in Fig. 4. The model obtains the best performance on emotions with less confusion, such as anticipation (78.01%, F1 score = 0.77) and joy (73.94%, F1 score = 0.71). While emotions with higher levels of confusion, such as anticipation and sadness, may have lower accuracy rates even if they perform well during training. For instance, the F1 score for anticipation reached 0.79 during training, but the validation F1 score was only 0.61. Similarly, sadness had an F1 score of 0.81 during training, but the validation F1 score was also 0.61. These results indicate that while the model may struggle to accurately identify these emotions, it still performs well overall. Therefore, the degree of confusion significantly affects the model's ability to discriminate emotions.

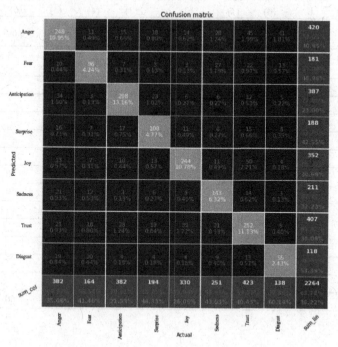

Fig. 4. The confusion matrix of analysis result

Implementation Result by Partial Redistribution. The result of the confusion matrix shows that the groups with higher confusion are: {Joy (4), Trust (6)}, {Anger (0), Trust (6)}, {Anger (0), Disgust (7)}, and {Anticipation (2), Trust (6)}. We performed the partial redistribution method to improve the overall F1 score of the model for the 28 sentiment groups in the order of error rate. The experimental results showed that 13 out of 28 sentiment groups were improved, and the results are summarized in Table 4. Since the partial redistribution method focuses on improving the discriminative ability of the model for sentiment groups that are prone to confusion. As a result, only the adjusted sentiment will experience a change in its F1 score. The second column, labeled "Overall," displays the validation accuracy to provide more detailed results. The other columns, each representing a specific emotion, display the F1 scores. The numbers in parentheses next to the emotion names represent the corresponding emotion numbers.

According to Table 4, three emotion groups obtained improved performance after applying the partial redistribution method, with validation accuracy rates over 64.2% for groups, such as {Anger (0), Surprise (3)}, {Anticipation (2), Trust (6)}, and {Surprise (3), Trust (6)}. While the partial redistribution method aims to improve the overall accuracy by addressing sentiment confusion, some groups showed improvement in only a single emotion while others showed a decrease in discrimination for another emotion, such as

Table 4. F1 score results table of the overall model and each emotion.

	Overall b	Anger (0)	Fear (1)	Anticipation (2)	Surprise (3)	Joy (4)	Sadness (5)	Trust (6)	Disgust (7)
Original result	0.638	0.61	0.55	0.77	0.56	0.71	0.61	0.60	0.42
Emotion Group	The results after using partial redistribution								
0,3	**0.642**	0.64	-	-	0.60	-	-	-	-
0,6	0.639	0.63	-	-	-	-	-	0.60	-
0,2	0.640	0.63	-	0.76	-	-	-	-	-
1,4	0.640	-	0.56	-	-	0.71	-	-	-
4,5	0.640	-	-	-	-	0.70	0.63	-	-
4,6	0.640	-	-	-	-	0.71	-	0.60	-
6,5	0.640	-	-	-	-	-	0.61	0.60	-
3,5	0.640	-	-	-	0.57	-	0.61	-	-
2,5	0.641	-	-	0.77	-	-	0.63	-	-
3,7	0.641	-	-	-	0.60	-	-	-	0.40
1,2	0.641	-	0.56	0.77	-	-	-	-	-
2,6	**0.644**	-	-	0.78	-	-	-	0.61	-
3,6	**0.644**	-	-	-	0.62	-	-	0.60	-

{Anger (0), Anticipation (2)}, {Joy (4), Sadness (5)}, and {Surprise (3), Disgust (7)}. These results suggest that there may be further factors that determine the impact of the partial redistribution method on the model's performance. We believe these are targets worthy of further investigation. In addition, in Sect. 3.3, we hypothesized that correcting groups with a high degree of confusion would lead to the greatest improvement in the model's overall performance. However, the results show that while the confusion level is related to the model accuracy, it is not necessarily positively related to the improvement in the model's performance after correction.

5 Conclusion

Sentiment analysis has become one of the major domains in SNS research. In addition to traditional polarity analysis, many studies focused on finer-grained SA to obtain more detailed analysis results for more immediate applications. However, the finer-grained classification represents a more complex definition of emotion features. For model discrimination, it is more likely to be confusing.

In this paper, we aimed to solve the confusion problem of SNS data in the SA process and proposed a partial redistribution method to improve the ambiguity in SA. This is an extension of the BERT-based model, which takes the results of BERT on SA, uses the confusion matrix to calculate the degree of confusion between emotions, and the data of emotional confusion groups are partially redistributed by the degree of confusion. Experimental results demonstrated that our proposed partial redistribution method could improve the confusion problem. The average classification accuracy has been improved to 64%.

However, there are still some limitations. For example, the partial redistribution method improves the overall accuracy, but it is possible that some sentiment discrimination rate decreases as a result. And improving the most confusing sentiment does not improve the performance of the model to the greatest extent. These problems deserve further study in our future work.

References

1. Hassan, A.U., Hussain, J., Hussain, M., Sadiq, M., Lee, S.: Sentiment analysis of social networking services data using machine learning approach for the measurement of depression. In: 2017 International Conference on Information and Communication Technology Convergence (ICTC), pp. 138–140. IEEE, Korea (2017)
2. Sharma, E., Gaur, A., Singhal, S.: Twitter sentiment analysis of India vs Pakistan T20 World Cup match using SVM classifier. In: 2022 International Conference on Machine Learning, Big Data, Cloud and Parallel Computing (COM-IT-CON), pp. 16–19. IEEE, India (2022)
3. Wu, D.D., Zheng, L., Olson, D.L.: A decision support approach for online stock forum sentiment analysis. IEEE Trans. Syst. Man Cybern. Syst. **44**(8), 1077–1087 (2014)
4. Agarwal,A., Yadav, A., Vishwakarma, D.K.: Multimodal sentiment analysis via RNN variants. In: 2019 IEEE International Conference on Big Data, Cloud Computing, Data Science and Engineering (BCD). pp. 19–23. IEEE, USA (2019)
5. Liao, S., Wang, J., Yu, R., Sato, K., Cheng, Z.: CNN for situations understanding based on sentiment analysis of twitter data. Proc. Comput. Sci. **111**, 376–381 (2017)
6. Liu, G., Guo, J.: Bidirectional LSTM with attention mechanism and convolutional layer for text classification. Neurocomputing **337**, 325–338 (2019)
7. Liu, C., et al.: Improving sentiment analysis accuracy with emoji embedding. J. Saf. Sci. Resil. **2**(4), 246–252 (2021)
8. Basiri, M.E., Nemati, S., Abdar, M., Cambria, E., Acharya, U.R.: An attention-based bidirectional CNN-RNN deep model for sentiment analysis. Futur. Gener. Comput. Syst. **115**, 279–294 (2021)
9. Devlin, J., Chang, M.W., Lee, K., Toutanova, K.: BERT: pre-training of deep bidirectional transformers for language understanding. arXiv:1810.04805 (2018)
10. Nafis, I.T., Mohammed, E.A.: Detecting multilabel sentiment and emotions from Bangla YouTube comments. In: 2018 International Conference on Bangla Speech and Language Processing (ICBSLP), pp. 1–6. IEEE, Bangladesh (2018)
11. Jin, Z., Lai, X., Cao, J.: Multi-label sentiment analysis base on BERT with modified TF-IDF. In: 2020 IEEE International Symposium on Product Compliance Engineering-Asia (ISPCE-CN), pp. 1–6. IEEE, China (2020)
12. Fei, H., Ji, D., Zhang, Y., Ren, Y.: Topic-enhanced capsule network for multi-label emotion classification. IEEE/ACM Trans. Audio, Speech, Lang. Process. **28**, 1839–1848 (2020)
13. Tang, T., Tang, X., Yuan, T.: Fine-tuning BERT for multi-label sentiment analysis in unbalanced code-switching text. IEEE Access **8**, 193248–193256 (2020)
14. Li, X., et al.: Enhancing BERT representation with context-aware embedding for aspect-based sentiment analysis. IEEE Access **8**, 46868–46876 (2020)
15. Song, Y., Wang, J., Liang, Z., Liu, Z., Jiang, T.: B Utilizing BERT Intermediate Layers for Aspect Based Sentiment Analysis and Natural Language Inference. arXiv:1810.04805 (2020)

16. Loshchilov, I., Hutter, F.: Decoupled weight decay regularization. arXiv:1711.05101 (2019)
17. Demszky, D., Movshovitz-Attias, D., Ko, J., Cowen, A., Nemade, G., Ravi, S.: GoEmotions: A Dataset of Fine-Grained Emotions. In: The 58th Annual Meeting of the Association for Computational Linguistics Proceedings, Association for Computational Linguistics, pp. 4040–4054. Association for Computational Linguistics, online (2020)
18. Plutchik, R.: The Nature of Emotions: Human emotions have deep evolutionary roots, a fact that may explain their complexity and provide tools for clinical practice 89(4), 344–350 (2001)

A Scientific Perspective of Agnihotra to Curtail Pollutants in the Air

Shailee Bhatia[1]([⊠]), Shelly Sachdeva[2], and Puneet Goswami[1]

[1] SRM University, Sonepat, India
shaileebhatia@gmail.com, goswamipuneet@gmail.com
[2] NIT, New Delhi, India
shellysachdeva@nitdelhi.ac.in

Abstract. Air Pollution is harmful to human health. Poor air quality can lead to life-threatening diseases. We have to defend ourselves from the viruses entering our immune system. Agnihotra/Yagya is a gift to humanity from ancient Vedic Science. It refers to a ritual performed through the fire prepared in a copper pyramid, along with the chanting of Sanskrit mantras. We aim to analyze the effect of Yagya on air pollutants. In our research, we propose a novel system using the Internet of Things (IoT), cloud technologies followed by data analytics to measure the concentration of pollutants in the air. The data will be gathered from IoT-based sensing units and stored on a cloud system. The analytical model is automated through machine learning and data visualization techniques. The pollutant concentration is measured before and after Agnihotra. The amalgamation of Vedic science with computer technology may lead to sustainable development by curtailing pollutants in the air.

Keywords: Agnihotra · Yagya · Air pollutants

1 Introduction

Agnihotra is an ancient Vedic ritual described in Yajurveda. Agni means fire and hotra means healing. Agnihotra is an integral part of Vedic culture, where a ritual is performed by igniting a sacred fire in a pyramid shaped pot.

Air quality is degrading worldwide, and according to World Health Organization data, more than 7 million people die due to air pollution alone. This global warming has now become a 'Global Warning'. There is an urgent need to continuously monitor the quality of air and use proper techniques to reduce pollution.

We intend to develop an energy-effective device which can gather information about the concentration of air pollutants and send this information to a cloud storage device to perform analysis. We also propose a solution, 'Agnihotra', that can reduce the concentration of pollutants in the air. The presence of particulate matter in the air significantly impacts one's health. Inhalation of excessive particulate matter can lead to various problems such as heart attack, asthma, and premature death. Particulate matter $PM_{2.5}$ and PM_{10} particles tend to settle on the ground or in water, contaminating the ecosystem due to their chemical composition.

S. Sachdeva et al. (Eds.): BDA 2022, LNCS 13830, pp. 211–219, 2023.
https://doi.org/10.1007/978-3-031-28350-5_17

We can make changes in the aura through Yagya. The vibrations created by chanting of Sanskrit mantras and the sacred fire prepared with organic substances during the sunrise and sunset time cycle have a very positive effect on the atmosphere. The fire emanating from Agnihotra disintegrates the Raja-Tama particles (impure subtle components) and creates a pure aura. An enormous amount of energy accumulates around the Agnihotra copper pyramid. It produces a magnetic field that balances off the negative energy and intensifies positive energy. Therefore, by continuing this process, the sacred fire creates a protective shield of up to 10 feet on all sides of the human being and protects the environment by reducing pollutants in the air.

2 State-of-the-Art

In recent years, several experiments have been performed by various researchers in the field of Agnihotra to reduce air pollution.

2.1 Balkrishna Acharya et al. (2022) described a traditional ayurvedic air sterilization technique, Dhoopan, to sterilize the air. The rooms fumigated with dhoopan for 30 min reduced the environmental microbial loads by ten times in real-world conditions using the agar plates [1].

2.2 Rastogi et al. (2021) talked about air quality analysis and used the conventional method 'Yagya'. The fire was prepared as per the process. Particulate Matter ($PM_{2.5}$, PM_{10}), temperature, humidity and carbon dioxide were measured with the help of the Air Veda instrument during and after Agnihotra [2].

2.3 Chaganti, Venkata R. (2020) described the process of Yagya. The study aimed to find out the effects of Agnihotra on air pollution. The Yagya experiments were conducted during the year 2012 to 2019 in the state of Georgia, USA to find the effect on the particulate matter before and after Yagya. All showed encouraging results about the effectiveness of Agnihotra as a solution to air pollution [3].

2.4 Yu-Lin Zhao et al. (2020) developed a system to monitor various air quality metrics from IoT devices and suggested an intelligent monitoring system to control air pollution [4].

2.5 Mahendra K. Gupta et al. (2019) focused on evaluating the effects of Agnihotra fumes and ash to make the contaminated air and water free from microbes. The settle plate method was used for passive monitoring of air samples [5]. The study showed encouraging results about the effectiveness of Agnihotra.

2.6 Roux et al. (2019) proposed a wet electrostatic precipitator method for air depollution systems over dense geographic areas with high pollution levels [6].

2.7 Shivhare and Gour (2019) conducted a study to explore the effects of Agnihotra in managing the environment and health. It is clear from the scientific studies that the Havan was designed to clean the environment [7]. The smoke of the Havan is not only used to disinfect the air but also helps in the physical, mental, intellectual and spiritual development of humankind. Havan was considered one of the most economical means of purifying environmental pollution.

2.8 Mamta et al. (2018) studied the effect of Yagya therapy on Particulate Matter. The $PM_{2.5}$, PM_{10}, CO_2, temperature and air concentrations were recorded using Digital PM

Sampler for one day before Yagya, on the day of Yagya Kriya and two days after Yagya. The study showed a reduction in $PM_{2.5}$, PM_{10} and CO_2 after performing Yagya in an indoor environment [8].

2.9 Abhang et al. (2015) studied the effect/result of Yagya fumes on the oxides of sulphur and nitrogen in the surroundings. Yagyas were performed in the two cities of Maharashtra. Air samples were collected using respective absorbing reagents for SO_x and NO_x with the help of a Handy sampler. SO_x level decreases, NO_x level increases initially but decreases afterward [9].

The concentration of pollutants is measured and analyzed, to examine the impact of Agnihotra on the atmosphere. The primary pollutant to determine air quality is particulate matter ($PM_{2.5}$ and PM_{10}). Some authors have also studied the impact of humidity, pathogenic bacteria, temperature, O_3, CH_4, SO_x and NO_x after and before the Yagya practice. Agar plates, Digital sampler, Settle plate method and Air Veda instruments were used by various authors to measure the concentration of pollutants in the air. In our research, IoT sensors will assist in monitoring the presence of pollutants in the surrounding area. They can be used for indoor and outdoor environments and offer monitoring at a lower cost than other conventional methods. Various studies indicate that Agnihotra has reduced the percentage of pollutants and helped purify the air [11].

3 Agnihotra Process

The process of Yagya purifies the atmosphere, replenishes oxygen and protects the ozone layer. It saves mankind from air borne diseases and helps protect our natural climate. The Agnihotra process is performed twice a day at sunrise and sunset.

Vedic Agnihotra involves preparing the fire in a copper pyramid shaped pot with dried cow dung and twigs of herbal trees. Pieces of dried cow dung dipped in ghee (clarified butter) are placed at the bottom of the pot in such a way as to allow air to pass through it [11]. A piece of camphor is placed in a spoon, and after lighting it with a match stick, it is placed in the havan kund (fire-pan), wherein the cow dung cakes are arranged. Offerings are made to the holy fire along with the chanting of Sanskrit mantras. We offer samidhas after the word "Swaha" to keep the divine fire ignited. Samidhas are mostly twigs that have holy significance according to Indian mythology. Sandal (Chandan), Banyan (Bargad), Mango (Aam), Peepal (Peepal) and Blackberry (Jamun) are considered holy. Ghee is the primary ingredient that serves as fuel to the divine fire. The substances viz. Cow dung, rice, ghee have antimicrobial and antioxidant properties; when burnt under appropriate conditions, combustion takes place which dissolves the toxic contents in the air. The heat energy generated from the holy fire and sound energy produced by reciting mantras emanates a powerful energy into the atmosphere. Figure 1 shows Agnihotra radiating into the atmosphere. The fumes produced are subject to the photochemical reaction of sunlight and undergo various modifications. Carbon dioxide in the atmosphere is also reduced to formaldehyde which acts as a powerful antiseptic and has a germicidal effect. The chemical reaction that takes place during the process of Agnihotra is as follows:

$$CO_2 + H_2O \;\rightarrow\; HCHO + O_2$$
$$\text{Carbon dioxide} + \text{water} \rightarrow \text{formaldehyde} + \text{oxygen}$$

(1)

The volatile substances produced from burning the cow's ghee from Yagya diffuse into the atmosphere. During Homa/Yagya process, the fumes that are liberated purifies the atmosphere with the help of some chemical combination in the surrounding air. Agnihotra gives positive energy to all living organisms.

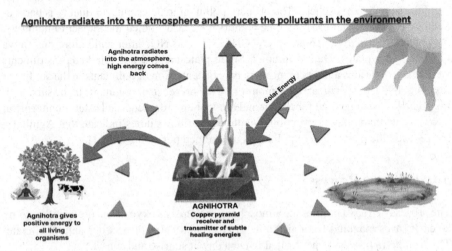

Fig. 1. Agnihotra process

4   Methodology

In this section, we give an outline of the methodology used. The proposed architecture diagram is depicted in Fig. 2. Data is collected from open sources and also through sensors deployed. The IoT device fosters the need of collecting & storing the concentration values of the major air pollutant (particulate matter). The analysis is done on the same to determine its accuracy by comparing the collected data from various open-source outlets and through sensors. The sensor measures the concentration of particulate matter in the air. Other meteorological aspects like humidity, temperature etc. are also collected periodically along with their timestamps. The server is facilitated at Firebase to gather information from sensors and open sources. Firestore interface is built for automatic scaling and ease of application development. The sensor is connected to an Arduino board which is an open-source firmware. The Arduino board can be easily configured using the Arduino IDE. A code is written to record the data and sending data from the serial monitor to ThingSpeak using python script and making HTTP requests to send the data. ThingSpeak is an IoT analytics platform service that allows to aggregate, visualize, and analyse data streams in the cloud. Pre-processing data involves data standardization, date-time indexing and joining and handling the missing values. Pollutants data has been fetched from the official website of the Central Pollution Control Board (CPCB) [10] with respect to the time frame at which the IoT device has recorded the data. The analytical server brings information from the data set server and performs further analysis. The application server utilizes the outcomes acquired from the examination and renders them to the application dashboard. The web application utilized for representation is facilitated on the server. The results are analyzed to examine the effect of Yagya on air pollutants.

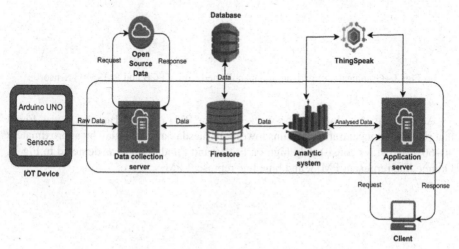

Fig. 2. Architecture diagram – data collection

5   Results

This section elucidates the graphical representation of the analysis done. Dataset collected at the server using the IoT setup is cleaned by removing null and erroneous values along with detecting and removing the outliers. Figure 3 depicts the graphical representation of the comparison of the data collected by the setup and the data collected from CPCB; it is evident that the sensor is accurately taking readings. The X axis is date and time, and the Y axis is pollutant concentration in μg/m3. The blue line represents the concentration values of PM_{10} recorded by the Central Pollution Control Board on a daily basis between 27th November 2021 to 20th January 2022. The dotted line depicts the concentration values of PM_{10} recorded using the IoT device with the Plant Tower PMS5003 Sensor. The recordings were taken twice daily for 15–20 min each and were recorded between 27th November 2021 to 20th January 2022.

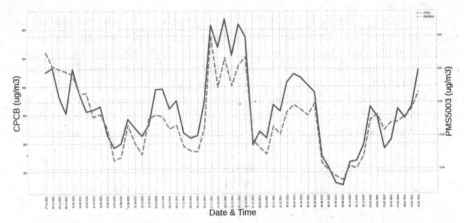

Fig. 3. Graphical comparison of data collected from CPCB and PMS5003 Sensor

An analysis was done on the week basis for $PM_{2.5}$ and PM_{10}, and it was found that pollutant concentration remains lowest on Tuesday and Saturday by approximately 40–60% and stays consistently high on Friday and Sunday night as depicted in Fig. 4. Orange line represent PM_{10}, and blue line represent $PM_{2.5}$.

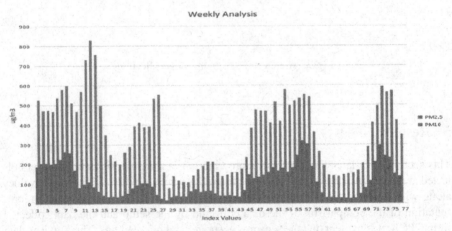

Fig. 4. PM_{10} and $PM_{2.5}$ weekly analysis

A day analysis was done on the pollutants PM_{10} and $PM_{2.5}$, and it was found that the concentration increases by 50% during 9–12 AM, and it keeps on dropping during the afternoon, and a slight increase during the night. The same is depicted in Fig. 5.

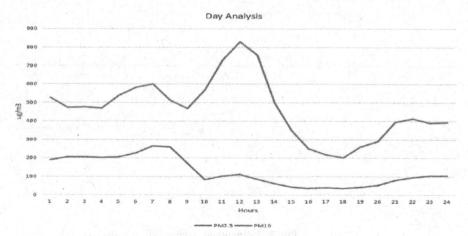

Fig. 5. PM_{10} and $PM_{2.5}$ daily analysis

The graphical analysis shown in Fig. 6 depicts the effect of Yagya on particulate matter PM_{10} and $PM_{2.5}$. The pollutant concentration has decreased rapidly when calculated in the Arya Samaj Mandir, Delhi, with around a 65% decrease in the concentration after performing Yagya. This infers that Yagya curtails the air pollutants.

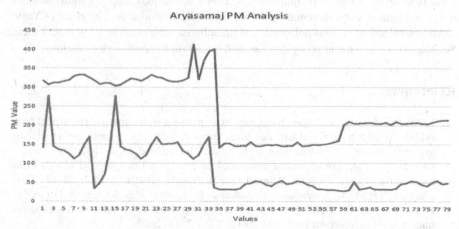

Fig. 6. PM_{10} and $PM_{2.5}$ analysis at Arya Samaj, Delhi

One interesting inference was that when humidity was measured, a sharp drop of 50% was seen in Arya Samaj Mandir, where the burning of twigs of herbal trees would kill the humidity, as shown in Fig. 7.

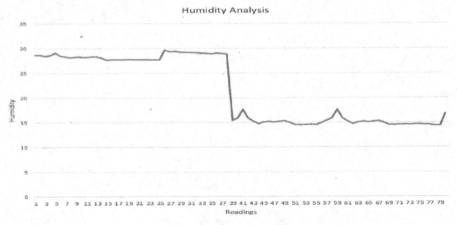

Fig. 7. Humidity analysis

6   Conclusion

Yagya is the need of the hour, a gift of Vedic science to humanity. The analysis presented in the result section shows that performing Agnihotra on a daily basis had a significant effect on reducing fine particulate matter in indoor air. Thus, a simple process of Agnihotra helps in reducing air pollution levels. This research may lead to a future solution to the problems of environmental pollution, especially air pollution. The effect of Yagya can be examined on various pollutants and greenhouse gases such as CO_2, methane etc. So, Yagya could be the next big thing around the world to counter air pollution.

References

1. Balkrishna, A., et al.: Vishaghn dhoop, nano-scale particles with detoxifying medicinal fume, exhibits robust anti-microbial activities: implications of disinfection potentials of a traditional ayurvedic air sterilization technique. J. Evid. Based Integr. Med. **27**, 1–15 (2022)
2. Rastogi, R., et al.: Computational analysis of air quality and the potential of rich Indian tradition for healthcare 4.0. Int. J. Reliab. Qual. E-Healthc. (IJRQEH). **10**(3), 32–52 (2021)
3. Chaganti, V.R.: Yajna A solution to air pollution. Int. J. Innov. Res. Sci. Eng. **8**, 11 (2020)
4. Zhao, Y.-L., et al.: Development of iot technologies for air pollution prevention and improvement. Aerosol Air Qual. Res. **20**(12), 2874–2888 (2020)
5. Rajpoot, S., Gupta, M.K.: Effect of Agnihotra on Microbial Load in Air Water and Genotoxicity of Onion Root Tips, vol. 8, pp. 2681–268 (2019)
6. Roux, J.-M., Achard, J.-L.: Wet electrostatic precipitators to reduce urban air pollution. In: 2019 International Conference on ENERGY and ENVIRONMENT (CIEM). IEEE (2019)
7. Shivhare, N., Gour, A.: Management of atmosphere and health using Hawan technique–a review. J. Current Sci. **20**, 1–7 (2019)
8. Mamta, S., Kumar, B., Matharu, S.: Impact of Yagya on particulate matters. Interdiscip. J. Yagya Res. **1**(1), 01–08 (2018)
9. Abhang, P., Pathade G.: Study the effects of Yadnya fumes on SOx and NOx levels in the surrounding environment. TATTVADIPAH, Res. J. Acad. Sanskrit Res. (2015)

10. CCR. (n.d.). https://app.cpcbccr.com/ccr/#/caaqm-dashboard-all/caaqm-landing/data. 30 Nov 2021
11. Bhatia, S., Sachdeva, S., Goswami, P.: Integrating IoT with Vedic Technology to Purify Atmosphere. In: 2021 3rd International Conference on Advances in Computing, Communication Control and Networking (ICAC3N). IEEE (2021)

Policy Driven Epidemiological (PDE) Model for Prediction of COVID-19 in India

Sakshi Gupta[✉] and Shikha Mehta

Computer Science and Engineering, Jaypee Institute of Information Technology, Noida, India
sakshi.officialid@gmail.com

Abstract. The fast spread of COVID-19 has made it a global issue. Despite various efforts, proper forecasting of COVID-19 spread is still in question. Government lockdown policies play a critical role in controlling the spread of coronavirus. However, existing prediction models have ignored lockdown policies and only focused on other features such as age, sex ratio, travel history, daily cases etc. This work proposes a Policy Driven Epidemiological (PDE) Model with Temporal, Structural, Profile, Policy and Interaction Features to forecast COVID-19 in India and its 6 states. PDE model integrates two models: Susceptible-Infected-Recovered-Deceased (SIRD) and Topical affinity propagation (TAP) model to predict the infection spread within a network for a given set of infected users. The performance of PDE model is assessed with respect to linear regression model, three epidemiological models (Susceptible-Infectious-Recovered-Model (SIR), Susceptible-Exposed-Infectious-Recovered-Model (SEIR) and SIRD) and two diffusion models (Time Constant Cascade Model and Time Decay Feature Cascade Model). Experimental evaluation for India and six Indian states with respect to different government policies from 15th June to 30th June, i.e., Maharashtra, Gujarat, Tamil Nadu, Delhi, Rajasthan and Uttar Pradesh divulge that prediction accuracy of PDE model is in close proximity with the real time for the considered time frame. Results illustrate that PDE model predicted the COVID-19 cases up to 94% accuracy and reduced the Normalize Mean Squared Error (NMSE) up to 50%, 35% and 42% with respect to linear regression, epidemiological models and diffusion models, respectively.

Keywords: COVID-19 · Coronavirus · India · Predictions · Forecasting · Epidemiological model

1 Introduction

The COVID-19 pandemic has been declared as global health issue as it is considered as the most harmful respiratory virus since 1918 H1N1 influenza [1]. It is an infectious disease that was primarily reported in Wuhan, China on 31st December 2019 [2]. By March 2020, COVID-19 was declared as pandemic by World Health Organization (WHO). The spread of COVID19 has been so fast that 215 countries have by now reported at-least one case. The COVID-19 report of WHO reveals a total of 17,748,367 confirmed cases and 681,974 deaths across the world by July 31, 2020. India reported its first corona virus in

© The Author(s), under exclusive license to Springer Nature Switzerland AG 2023
S. Sachdeva et al. (Eds.): BDA 2022, LNCS 13830, pp. 220–243, 2023.
https://doi.org/10.1007/978-3-031-28350-5_18

January 2020 and by July; there were 1,697,054 cases and 36,551 deaths in India. Being second largest populated country, whole world is keeping an eye on the policies adopted by Indian government to control the spread of epidemic as shown in Fig. 1.

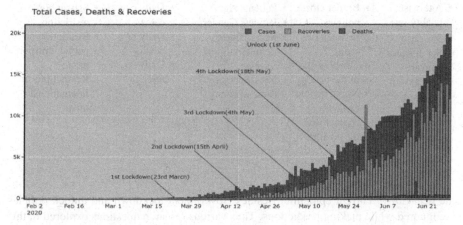

Fig. 1. Total cases, death, recoveries in India at various stages of lockdown

With so much population and limitations of available resources, the in-depth study of COVID-19 outbreak in Indian region is critical. This derives the need of systematic study the impact of various strategies adopted by Indian government. In this respect, various studies have been performed to analyze and forecast the coronavirus. In literature, researchers have used the various types of data provided by government to study the behavior of the epidemic and its impact in controlling the spread rate of COVID-19 [10]. The various features of data used to analyze the epidemic impact over the globe by various researchers are daily new cases, daily new deaths, age-wise fatality rate, patient specific information, stay home restrictions, lockdown policies. In this work, all these features are classified into five types: policy features, temporal features, interaction features, structural features and profile features and as shown in Table 1.

Some of the existing studies have used these features to study the spread patterns of COVID-19. Jiang et al. estimated that the fatality rate of COVID-19 is 8.0% for the age group 70–79, 14.8% for the age group >80 and 4.5% for the remaining people [5]. Hence, the persons of the age >50 and suffering from the diseases like Parkinson's disease, diabetes and cardiovascular disease to be considered at the highest risk. Overall, they have used the profile and structural features to study the COVID-19 impact. According to the World Health Organization report, profile features play a very important role to identify the severity of the disease. The common symptoms for this disease are fever, cough, shortness of breath which take 2–14 days to appear that can lead to pneumonia, kidney failure and even death which is highly depends upon the patient's medical history [6]. Wu et al. observed that the disease spread is also possible due to the respiratory droplets if people come in close contact with infected person [7]. On an average a COVID-19 patient may infect 1.5 - 3.5 persons considering that the virus is airborne. These studies reveal that most of the researchers have only considered one or two features to analyze

Table 1. Classification of features in epidemic analysis

Structural features	Interaction features	Profile features	Temporal features	Policy features
• Age-wise fatality rate • Sex ratio • Set of states who are neighbor of the state	• Border closer policies between two states • Travel history of the patient • No. of people came in contact with the positive case	• Patients Age • Patients Gender • Recent domestic/international traveler • No. of persons came in contact	• Daily new cases • Daily new deaths • Daily no. of tests	• Stay at home restriction • Restriction on public transport usage • Restrictions on international travel

the behavior of current epidemic. Besides Government lockdown and unlock policies are ignored while making predictions. Thus various research questions explored in this paper are:

- What is the role of the different policies/strategies implemented by Indian government to handle the outbreak of COVID19 epidemic?
- Whether a particular state will be able to handle the critical coronavirus situation by predicting the future number of coronavirus patients in that particular state based on the previous trends and their imposed policies?
- What is the influence of structural, profile, interaction, profile and policy features on COVID-19 spread?

Literature reveals that machine learning algorithms and mathematical models provide better predictions in healthcare [1, 17, 46] problems with large amount of data. The comprehensive review by Nsoesie et al. has presented various techniques employed for the analysis of influenza pandemic such as Regression models, prediction rules, Bayesian network Susceptible-Infectious-Recovered-Model (SIR), Susceptible-Exposed-Infectious-Recovered-Model (SEIR) and Susceptible-Infectious-Recovered-Deceased-Model (SIRD) [8]. However, these studies of COVID-19 are limited to exploratory analysis with small amount of data [9].

Since, coronavirus is a detrimental respiratory, i.e., an individual can be infected by the other infected individuals and this process continues which makes the spread of virus diffusive. Diffusion is a process of the spreading of something more widely. Therefore, in this work, a diffusion model is proposed which predicts the number of daily cases while considering the temporal, structural, interaction and profile features of the individual and the group such as Indian government imposed public policies designed to handle this epidemic, available medical facilities, the number of reported cases, deaths and cases that have recovered, and characteristics of the individuals. Proposed model can be used to project the timeline of the new corona-positive cases forward in time.

The objectives of the work are as follows.

- Develop a Policy Driven Epidemiological (PDE) model using temporal, interaction, structural, profile and policy features of individuals to Predict/Forecast COVID-19 in India.
- Incorporate the various constraints such as available medical facilities, population or age/gender ratio in the spread of the coronavirus.
- Analyze the role of the Indian government imposed public policies designed to handle this epidemic.
- Evaluated the performance of the proposed model against the various other contemporary model i.e., Linear regression, SIR, SEIR, SIRD, TC-C, and TDF-C models.

To the best of our knowledge, no work has been done to forecast the COVID-19 cases of all Indian states with respect to different government strategies. This work mainly focuses on Indian States and India as a whole. The rest of the paper is arranged as follows. Section 2 reviews the related work in the field of prediction and analysis of COVID-19. Section 3 presents the contemporary diffusion models in detail. Proposed model used for the analysis and prediction of the virus is explained in Sect. 4. After this, the dataset and the evaluation parameters are discussed in Sect. 5. Section 6 covers experiments, analysis and performance evaluation followed by conclusion in Sect. 7.

2 Related Work

The unavailability of Covid19 vaccine makes it imperative to manage the pandemic by controlling the epidemic peak that is flattening the epidemic curve. Data scientists across the globe are making efforts to apply technologies over available data to understand the spread behavior of virus to get the bigger picture and help the decisions makers. Various studies have been done to model and predict the spread size and end phase the epidemic. Wu et al. [11] developed a susceptible exposed infectious recovered (SEIR) meta-population model to understand and analyze the epidemic in China. Funk et al. presented a stochastic transmission dynamic model to study the virus transmission within and outside Wuhan [12]. Similarly, SEIR model is applied to study the impact of epidemic in Japan [13] and India [15]. Abdullah et al. [14] developed a stochastic SIR model for predicting the COVID-19 spread in Kuwait. Another work was done by Ndairou et al. [16], which mainly focused on the ability of super-spreader individuals to transmit the virus. Wang et al. [18] presented a Patient Information Based Algorithm (PIBA) to predict the death rate in Chine due to COVID-19. Gupta et al. performed the spread analysis on the United States and observed that there spread of the COVID-19 depends upon the temperature [19]. From the studies they predicted that the virus will become weak in the summer season and infected cases in India may go down in May in India which actually didn't happen. Ceylan has applied the ARIMA models for forecasting rise in COVID-19 cases in Italy, Spain and France with 4% to 6% of Mean absolute percentage error [21]. Similarly, Fanelli et al. used SIRD model for forecasting of demand in ventilation units due to COVID-19 in China, France and Italy [22]. Reddy et al. used long short term memory (LSTM) model to predict the end date of COVID-19 in Canada [23]. Arora et al. combined a recurrent neural network (RNN) model with long short-term memory (LSTM) to predict number of infected cases [17]. Table 2 depicts the various types of features used in the models developed so far for prediction of COVID-19.

It can be precisely observed from Table 2 that most of the studies have employed profile, structural and temporal features for COVID 19 for prediction of outbreak. Government policies which have been implemented time to time to control the spread of virus across the nation have been left untouched by the researchers. This work explores the impact of government policies along with structural, interaction, temporal and profile features to predict the spread of coronavirus.

3 Background

This section concisely presents the basic concepts of the contemporary models that have been used for the forecasting of the coronavirus spread. The models used for the forecasting or prediction of data are classified into three categories: Regression model, Epidemiological models and Diffusion models. Regression Model is one of the oldest and core methods in data sciences. Epidemiological models are used for the prediction of disease spread such as Susceptible-Infectious-Recovered-Model (SIR) [41]. Few other extensions have been amended in SIR are Susceptible-Exposed-Infectious-Recovered-Model (SEIR) and Susceptible-Infectious-Recovered-Deceased-Model (SIRD). Similarly, for the forecasting of diffusion in a network Time Constant Cascade Model [43] and Time Decay Feature Cascade Model [44] are applied in literature which are the originated from Independent Cascade model [42].

Linear Regression: It is supervised machine learning technique used for predicting the continuous data. It plots a straight line using the least squares method to minimize the distances between the predicted values and the resulting trend line.

Time Constant Cascade Model (TC-C): This model considers time as constant entity. The infection probability $P_t(i,j)_{TC-C}$ of any link e_{ij} does not depend on time t which indicates that for each timestamp t ($t \leq t + 1$) after state 'i' takes an action, $P_{t+1}(i,j)_{TC-C}$ remains the same as $P_t(i,j)_{TC-C}$.

Time Decay Feature Cascade Model (TDF-C): This model considers decay in time as an important feature. In some scenarios like lockdown when people interact either for a small duration or there is no interaction at all, the numbers of interactions decrease fast. This model analyzes the numbers of interactions over a period time to determine the link infection probability, i.e., $P_n(i,j)_{TDF-C}$. In this work, time period of interaction is divided into slots where, each slot is of 1-day time t.

Susceptible-Infectious-Recovered-Model (SIR): It is an epidemiological model that theoretically calculates the count of cases infected with a contagious illness for a closed population over a period time. The name of this model is derived based on the various equations used for computing the count of susceptible people S(t), infected people I(t), and the number of people who have recovered R(t).

Table 2. Features used in literature for COVID-19 spread analysis

Model name	Author [Reference]	Features applied	Type of data	Objective
Linear regression	Pandey et al. [1]	Structural, Temporal	Time series data	• For predicting outbreak of COVID-19 in India
Random Forest	Cobbet al. [24] Shi et al. [25] Tanget al. [26] Sarkar & Chakrabarti [27] Chen et al. [28]	Structural	Text data & Chest CT images	• Study the impact of social distancing • Analyze the predictors and their effects on mortality rate • Identification of factors negatively influencing the viral clearance
SVM	Sonbhadra et al. [29] Zhang et al. [30] Hassanien et al. [31] Barstugan et al. [32] Sethy et al. [33]	Structural, Profile	Text data, CT images and X-ray image	• Identification activities and trends inCOVID-19 research articles • Detection of severely ill patients based on mild symptoms • Predicting the recovery rate • For the classification of COVID-19 • To detect Coronavirus disease
SIR	Roda et al. [34]	Structural, Temporal	Time series data	• For predicting outburst of COVID-19

(continued)

Table 2. (*continued*)

Model name	Author [Reference]	Features applied	Type of data	Objective
SEIR	Wu et al. [11]		Time series data	• For prediction of COVID-19 spread in real time
SIRD	Fanelli et al. [22]	Structural, Temporal	Time series data	• Forecasting the survival rate of sternly ill cases • Prediction of mortality risk in severe infected COVID-19 patients
Machine learning models	Liu et al. [35] Dandekar et al. [36] Carrillo-Larco et al. [37] Magar et al. [38] Yan et al. [39] Chen et al. [40]	Structural, Profile, Temporal	Text data, Time series data and CT images	• For prediction of COVID-19 spread inreal time • Forecasting COVID-19 spread in Wuhan, China • Categorization of countries based on count of confirmed cases • Prediction of inhibitory synthetic antibodies of Corona virus • Mortality rate of critically infected patients

Susceptible-Exposed-Infectious-Recovered-Model (SEIR): In large number of infectious diseases, virus incubates within the body of hosts for a certain period of time before actually infecting the hosts. In this duration, an individualis said to be in compartment E (or exposed state). Thus, this model extends the SIR model to assess the impact of incubation in disease transmission, i.e., Susceptible - Exposed - Infectious - Recovered – Susceptible (SEIR) model [11].

Susceptible-Infectious-Recovered-Deceased-Model (SIRD): One of the popular variants of the SIR model is the Susceptible-Infectious-Recovered-Deceased (SIRD) Model. A preliminary analysis (SIRD) model is used quantify the epidemic. SIRD model distinguishes Recovered individuals that are the people who survived the disease and are now immune with respect to Deceased. The total population N of SIRD model is computed as sum total of susceptible (S), infected (I), recovered (R) and dead (D) for all times t. Hence $N = S + I + R + D$.

4 Policy Driven Epidemiological (PDE) Model

The proposed Policy Driven epidemiological (PDE) model is the integration of two models, i.e., Susceptible-Infected-Recovered-Deceased (SIRD) and Topical affinity propagation (TAP), to forecast the infection spread within a network for a given set of infected users. The process flow of the proposed PDE model is shown in Fig. 2. The objective is to develop a model that can efficiently forecast the virus spread using all five features (structural, temporal, profile, interaction, and policy), as discussed in Sect. 1.

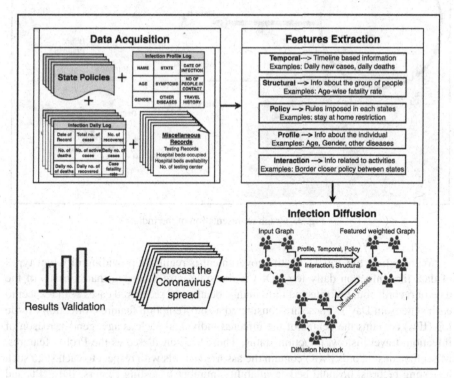

Fig. 2. Framework of the proposed policy driven epidemiological (PDE) model

The complete working of the PDE model is alienated into three steps. The first step is data acquisition, and the second step is feature extraction. The last step of PDE is the infection diffusion. The complete details of these steps are as follows:

4.1 Data Acquisition and Feature Extraction

A graph $G = \{V, E\}$ represents the map of India as shown in Fig. 3. Where V is the set of states of the country (represents the nodes in the graph), E is the set of links, i.e., e_{AB} means node 'A' and node 'B' are neighboring states as 'A = Delhi' and 'B = Uttar Pradesh'. In this paper, graph G is considered as the weighted networks such that each state 'A' carries a weight w_A. The epidemiological weight w_A is the infection susceptible probability of state 'A' of the graph G. Each state 'A' is also a graph, i.e., state-network $S_A = (P_A, C_A)$. Where, P_A refers to the set of persons in state 'A', and C_A is the set of connections between the persons, i.e., connection between persons, i.e., c_{Aij} represents person 'i' is connected with person 'j'. Based on whether the infection has reached the user or not, the status of each person in the network is distinguished into one of the four categories- infected, recovered/immune, susceptible, Dead.

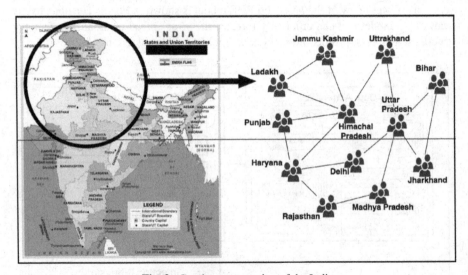

Fig. 3. Graph representation of the India

With respect to COVID-19, the government is regularly providing different types of data like Infection daily log (IDL) is the state-wise daily log that consists of the date of record, count of infected individuals, deaths and recovered cases with respect to each state from Day 1. These are considered as the Temporal features. Infection Profile log (IPL) contains the details of the infected individual such as age, gender, reason of infection, travel history, infection status. These are considered as the Profile features. Miscellaneous records (MR) contain the assorted details with respect to each state such as testing records, hospital beds availability, number of testing centers. IDL, IPL and MR are used altogether to extract the Structural features. State policy log (SPL) contains the details about the all 16 policies (as shown in Table 3) with respect to each state. Each policy value is categorized into one of the three stages, i.e., (No restriction, Partial restriction or Full restriction). These are used to extract the Interactions features. The value is different for each state and varies from time to time.

Therefore, State policy log (SPL) contains the policy imposed in each state from Day 1 and calculates the daily policy score with respect to each state. Policy score calculation is described in Eq. 1.

$$(Policy\ Score)_A = \frac{1}{k}\sum\nolimits_{i=1}^{k} P_{Aj} \tag{1}$$

where k is the number of policies imposed in the state and P_{Aj} is the sub-index score for an individual policy with respect to state 'A'.

From IPL, SPL and IDL, graph G is converted into the featured graph. The complete set of features with respect to every state is computed, i.e., Feature_set (F). Where, f_A represents the feature vector of state 'A' and stores the value of n different features (Temporal, structural, profile, policy and interaction) of state 'A' as $f_a = \{f_{a1}, f_{a2}, f_{a3}, f_{a4}, ..., f_{an}\}$ as shown in Fig. 2.

Table 3. List of policies

Policy ID	Name	Description	Applicability	Rules
P1	Restriction on Educational field	Closings of schools and universities	Country/State-wise	0 - No restriction/data unavailable 1 - Partial restriction (only some levels or categories, such as open for teaching staff and closed for students) 2 - Full restriction
P2	Restriction on Workplace closing	Closings of workplaces/Industries	Country/State-wise	0 - No restriction/ data unavailable 1 - Partial restriction (only some levels or categories, such as open of some sectors (banking sector, grocery stores) and Work from Home for remaining) 2 – Full restriction (Work from Home for all)

(continued)

Table 3. (*continued*)

Policy ID	Name	Description	Applicability	Rules
P3	Restrictions on public events	Cancellation of public events	Country/State-wise	0 - No restriction/ data unavailable 1 - Partial restriction (allowed with up 50 people gathering/ essential events like blood donation camp or food distribution camps) 2 – Full restriction
P4	Restrictions on private functions/ceremonies	Limits on private gatherings	Country/State-wise	0 - No restriction/ data unavailable 1 - Partial restriction (only 0–50 people are allowed to attend the ceremonies) 2 – Full restriction (restriction on gathering of 4 people or more)
P5	Restriction on public transport usage	Closing of public transport	Country/State-wise	0 - No restriction/ data unavailable 1 - Partial restriction (significantly reduce capacity of available means of transport) 2 – Full restriction

(*continued*)

Table 3. (*continued*)

Policy ID	Name	Description	Applicability	Rules
P6	Stay at home restriction	Orders to go out from home in case of emergency and otherwise stay at home	Country/State-wise	0 - No restriction/ data unavailable 1 - Partial restriction (people with special passes are allowed only) 2 – Full restriction
P7	Restrictions on internaltravel	Restrictions on travel between cities	Country/State-wise	0 - No restriction/ data unavailable 1 - Partial restriction (people with special permissions are allowed only) 2 – Full restriction
P8	Restrictions on internationaltravel	Restrictions on travel to other countries	Country/State-wise	0 - No restriction/ data unavailable 1 - Partial restriction(people with special permissions are allowed only) 2 – Full restriction
P9	Basic living support	The government is providing direct cash payments/ food	Country/State-wise	0 - No support/ data unavailable 1 - Partial support (provided to people with below poverty income) 2 – Full support

(*continued*)

Table 3. (*continued*)

Policy ID	Name	Description	Applicability	Rules
P10	Financial Support	The government is giving relaxation in financial obligations (eg stopping loan repayments, electricity or water bills)	Country/State-wise	0 - No support/ data unavailable 1 - Partial support (in the specific fields only such as income tax filling) 2 – Full support
P11	Use of Aarogyasetu mobile app	Entry restrictions based on the Aarogyasetu app	Country/State-wise	0 - No restriction/ data unavailable 1 - Partial support (mandatory to have in mobile to access specific places such as workplace) 2 – Full restriction
P12	Restrictions on Testing	The government policy specifies on who has access to COVID-19 testing	Country/State-wise	0 - No restriction/ data unavailable 1 - Partial restriction (testing of anyone showing Covid-19 symptoms) 2 – Full restriction (only those who meet two criteria (a) have symptoms (b) They are key worker/ admitted to hospital/came into direct contact with a known case/ returned from overseas)

(*continued*)

Table 3. (*continued*)

Policy ID	Name	Description	Applicability	Rules
P13	Contact tracing	The contact tracing policy of government after a positive diagnosis	Country/State-wise	0 - No restriction/ data unavailable 1 - Partial restriction (14 days of home-isolation is advised to all traced people) 2 – Full restriction (compulsory COVID-19 testing and home isolation of 14 days to all traced people)

4.2 Epidemiological Weight Computation

The epidemiological weight w_A is the infection susceptible probability of each state 'A' of the graph G that shows by what probability the infection will spread in the state 'A'. In this paper, epidemiological weight w_A is computed using the extended topical affinity propagation method [20]. i.e., E-TAP. E-TAP algorithm is able to calculate the k different weight values $\{w^1_A, w^2_A,..., w^k_A\}$ with respect to each state 'A' of the graph G. Where k represents the number of features in the graph. In this paper, five specific features are considered:

- First feature is policy score which is calculated using 13 different policies imposed by each state;
- Second is profile features of the person such as number of people who came in contact with COVID-19-positive patient;
- Third is structural features such as medical facility status of each state;
- Fourth is temporal feature such as daily reported number of new cases, new deaths;
- Fifth is interaction features such as boarder closer restriction between states.

Therefore, $\{w^{policy\_score}_A, w^{temporal}_A, w^{structural}_A, w^{profile}_A, w^{interaction}_A\}$ is calculated with respect to each state 'A' of the graph G using TAP algorithm. The detailed description of the TAP algorithm is as follows:

Primarily, the feature_set F of the network G is given as input to E-TAP. Thereafter, c_wt for each state-paire$_{AB}$ in network G is computed using the Jaccard Index which is defined in Eq. 2. Jaccard Index computes the similarity between two sets with respect to each feature k, i.e., the number of features that are common in both sets divided by the total number of features in both the sets feature k. The mathematical representation of the c_wt of state-paire$_{AB}$ for feature k is as follows:

$$c\_wt_{ABk} = \frac{|f_{Ak} \cap f_{Bk}|}{|f_{Ak}| + |f_{Bk}| - |f_{Ak} \cap f_{Bk}|} \tag{2}$$

where the c_wt$_{ABk}$ is the Jaccard index of state-paire$_{AB}$ which is calculated with respect to each feature k, i.e., profile, interaction, temporal, policy and structural features (Fig. 2). Table 4 shows the various symbols used in this paper.

Table 4. List of symbols

Symbol	Description
G	Graph
U	Set of states
E	Set of edges
N_U	$|U|$
N_E	$|E|$
A	Symbolizes a State in Graph G
NB(A)	Set of neighbors of State 'A'
S_A	State network of State 'A'
(Policy Score)$_A$	Daily policy score of State 'A'
e_{AB}	Represents States A and State 'B' are neighbor
c_wt$_{ABk}$	Connection weight of the state-pair e_{AB} for feature k
m_{Ak}	State A's representative for feature k
L_{ABk}	Logarithm of normalized collective principal_state_function of e_{AB} w.r.t. feature k
Q_{ABk}	Probability of infection State 'B' suspected to coming from State 'A' w.r.t. feature k
O_{ABk}	Probability of infection State 'B' actually received form State 'A' w.r.t. feature k
Y_{ABk}	Infection probability of state-pair e_{AB} w.r.t feature k
$w^k{}_A$	Epidemiological weight of State 'A' w.r.t. feature k
W	Set of the epidemiological weight w_A corresponding to each state
Status	State of the person(infected/recovered/susceptible/dead)
I(t)	I(t) represents the number of people infected
D(t)	D(t) is the number of people deceased
R(t)	R(t) represent the number of people recovered
R0	Reproduction rate
Score(t)	Infection spread size

After computing the c_wt of with respect to each state-pair, the principal_state_function $d(A, m_{Ak})$ is calculated using Eq. 3

$$d(A, m_{Ak}) = \begin{cases} \dfrac{c\_wt_{Am_{Ak}k}}{\sum_{B \in NB(A)}(c\_wt_{ABk}+c\_wt_{BAk})} m_{Ak} \neq A \\ \dfrac{\sum_{B \in NB(A)}(c\_wt_{BAk})}{\sum_{B \in NB(A)}(c\_wt_{ABk}+c\_wt_{BAk})} m_{Ak} = A \end{cases} \quad (3)$$

where, m_{Ak} represents the state agent for State 'A' such that $m_{Ak} \in \{NB(A) \cup A\}$ with highest c_wt_total out of all neighbor for feature k i.e.

$$c\_wt\_total(A, k) = \sum_{C \in NB(A)} c\_wt_{ACk} \quad (4)$$

Using principal_state_function, the L_{ABk} is calculated. L_{ABk} is the logarithm of normalized collective principal_state_function of state-paire$_{AB}$ defined by Eq. 5.

$$L_{ABk} = \log \frac{d(A, m_{Ak}, k)|_{m_{Ak}=B}}{\sum_{C \in NB(A) \cup \{A\}} d(A, m_{Ak}, k)|_{m_{Ak}=C}} \quad (5)$$

The output value of L_{ABk} is used to calculate of O_{ABk}, Q_{BBk}, Q_{ABk}, and Y_{ABk} using Eqs. 6, 7, 8, and 9, respectively.

$$O_{ABk} = L_{ABk} - \max_{C \in NB(B)}\{L_{ACk} + Q_{ACk}\} \quad (6)$$

$$Q_{BBk} = \max_{C \in NB(B)} \min\{O_{CBk}, 0\} \quad (7)$$

$$Q_{ABk} = \min(\max\{O_{ABk}, 0\} - \min\{Q_{BBk}, 0\} - \max_{C \in NB(B)\setminus\{A\}}\min\{O_{CBk}, 0\}), \ A \in NB(B) \quad (8)$$

$$Y_{ABk} = \frac{1}{1 + e^{-(O_{BAk}+Q_{BAk})}} \quad (9)$$

Using the value of Y_{ABk} for each state-paire$_{AB}$ on feature k, epidemiological weight w_A is calculated with respect to each state using Eq. 10

$$w_A^k = \sum_{C \in NB(A)} Y_{CAk} \quad (10)$$

Using Algorithm 1, $\{w^{policy\_score}_A, w^{temporal}_A, w^{structural}_A, w^{profile}_A, w^{interaction}_A\}$ is calculated. Overall, infection rate of each state w_A is calculated using Eq. 11

$$w_A = \sum_k w_A^k \quad (11)$$

Overall, the set of estimated epidemiological weight w_A corresponding to every state of the network G is computed using Algorithm 1.

Algorithm 1: E-TAP (G, F, A)

1.	For each topic k
2.	Compute the principal_state_function using eq. 3
3.	Calculate L_{ABk} by eq. 5
4.	Assign all $Q_{ABk} = 0$
5.	Repeat until convergence
6.	For each state-paire$_{AB}$ in G
7.	Determine the value of O_{ABk} by eq. 6
8.	For each state A in G
9.	Determine the value of Q_{AAk} by eq. 7
10.	For each state-paire$_{AB}$ in G
11.	Determine the Q_{ABk} using eq. 8
12.	For each state A in G
13.	For each state $C \in NB(A) \cup \{A\}$
14.	Determine the value of Y_{CAk} by eq. 9
15.	For each state A in G
16.	Determine epidemiological weight w^k_A by eq. 10
17.	Determine epidemiological weight w_A by eq. 11
18.	Return G (U, E, W)

Therefore, the E-TAP calculated epidemiological weight w_A of each state 'A' using all five features, i.e., temporal, structural, profile, policy and interaction.

4.3 Infection Diffusion

After obtaining the featured graph and epidemiological weight, the next step is infection diffusion. This step discovers the diffusion effect of the infection in the network, i.e., how many users are getting infection between the timestamp t to t + 1. For this, epidemiological weight w_A is used which is computed with respect to each state 'A' of the graph G, and status of all persons is updated with respect to each state-network 'A' of the graph G.

The status of each person in the graph G is distinguished into one of the four categories-susceptible (S), infected (I), recovered (R) and dead (D). Thus, |G| = S + I + R + D. Whenever a susceptible person comes in contact with a communicable person, the susceptible may turn into infected. Literature suggests, an infected individual is not contagious or recovered after 3–6 weeks [45]. On the average, infectiousness begins to appear from 2–3 days before the symptoms are visible. Therefore, an infected individual can infect others before clear visibility of symptoms. The average time period to remain an infected person infectious is 12–14 days [45]. It has been observed in literature that the number of days from infection to death to be around 17.3 [3]. That means the person who died on 13th June 2020 probably got infected around 27th May 2020. Therefore, the number of deaths in a region can be used to guess the number of realistic current cases. Similarly, the estimated mortality rate is 7% [3]. That means that, around 27th May 2020, there were already around ~14.2x cases (of which only one ended up in death 17.3 days later). In this work, age distributions which is obtained from the Population Pyramid website [4], mortality rate of 7%, doubling rate of cases as 14 days [45], number of days from infection to death to be around 17.3 is taken to forecast the COVID-19 impact[3], and the numeric values of the model parameters are obtained through parameter tuning. Initial value of the number of susceptible, infected, recovered and deaths at time $t = 0$ are extracted from Infection daily log (IDL) and Infection Profile log (IPL). After this, to analyze and forecast the virus spread in the graph G such that how many people will become Covid-19 positive at timestamp $t + 1$, the following set of differential equations are used to compute the temporal dynamics of the population of susceptible (S), infected (I), recovered (R), dead (D):

$$\frac{dS(t)}{dt} = -\beta \frac{SI}{N} \tag{12}$$

$$\frac{dI(t)}{dt} = \beta \frac{SI}{N} - \gamma I - \delta I \tag{13}$$

$$\frac{dR(t)}{dt} = \gamma I \tag{14}$$

$$\frac{dD(t)}{dt} = \delta I \tag{15}$$

where t refers to time, S(t) represents count of susceptible people, I(t) is count of infected people, R(t) refers to population who have recovered and are now immune to the infection, D(t) is the count of deceased people, beta, gamma and delta represent infection rate, recovery rate, and death rate respectively. To evaluate the speed with which a human carrier would infect the set of susceptible people is computed as there production rate (R_0). The value of $R_0 > 1$ means that disease has begun to spread in the population infecting more people whereas diseases does not spread if $R_0 < 1$. Therefore, forecasting of spread majorly depends upon reproduction rate. In this work, epidemiological weight of state is used as the reproduction rate that determines the transmission speed of the pandemic in each state, i.e.,

$$W_A = R_0 = \frac{\beta}{\gamma} \tag{16}$$

Overall, the one iteration of the PDE model includes three steps, i.e., data acquisition, feature extraction, and infection diffusion. This way, the infection diffusion process runs for each state of the graph G and the infection spread Score(t) is computed as described in Eq. 13. The changes identified at each timestamp is added to the Infection daily log (IDL) and Infection Profile log (IPL), such that these predictions are utilized to improve infection forecast in the next iterations of the PDE model.

5 Evaluative Parameters

The efficacy of the proposed PDE model is assessed through analysis of the forecasting of the epidemic. The number of daily cases predicted by the PDE model is evaluated using the following parameter:

- **Predicting Spread:** The total number of individuals estimated to get infected in the future is called the infection spread size [43]. The variation between the actual infection spread size at time t **Score$_A$(t)** and predicted infection spread size at time t **Score(t)** is known as the Normalized Mean Squared Error (NMSE) of the prediction size.

$$NMSE = \frac{1}{|V|} * \frac{(Score(t) - Score_A(t))^2}{Score(t)} \qquad (17)$$

6 Experiments and Analysis

Thorough study was performed for evaluating the efficiency of the proposed PDE model for infection diffusion as compared to other contemporary model i.e., Linear regression, SIR, SEIR, SIRD, TC-C, and TDF-C models. All algorithms were implemented in Python version 2.7 using macOS operating system with 8GB RAM and 1.8GHz Intel Core i5 processor.

6.1 Datasets

To answer the different research questions identified in this work, following information is extracted with respect to India from the various resources for the period 31st December 2019 to 30th June 2020:

1. Age Group details – AgeGroup, Total Cases, Percentage.
2. Hospital Beds – State, No of health centers, hospital and beds in that state.
3. ICMR Testing Labs – Name of the lab, Address, City, State, Type (Government or private)
4. Individual Details – Diagnosed date, Age, Gender, Detected State, Current Status. Details were available for only 250000 patients.
5. State wise Testing Details – Date, State, no of samples tested, no of positives, negatives on that particular date.
6. COVID-19 India – State, Date, Confirmed Cases, Cured and deaths on that date.
7. Population of India – State, Population (Rural and Urban), Area, Density, Gender Ratio.

The data from period 31/12/2019 to 15/06/2020 is used as the training data and data from 16/06/2020 to 30/06/202 is used as the testing data.

6.2   Analysis and Discussion of Results

The performance of the models is evaluated for India and six Indian states with respect to different government policies from 15th June to 30th June, i.e., Gujarat, Delhi, Rajasthan, Tamil Nadu, Maharashtra and Uttar Pradesh as follows:

- **COVID-19 Prediction Analysis:**

The performance of the PDE model is evaluated with respect to the various other contemporary models, i.e., Linear regression, SIR, SEIR, SIRD, TC-C, and TDF-C models for the period of 15 days from 15th June to 30th June 2020. This experiment evaluates the prediction of the infection spread size **Score$_A$(t)**, that is predicting the number of infected users at any given time t in state 'A'. This is considered as the direct measure of how well the predictions fit and what actually happened. Figure 4 depicts the results of all seven models for India. For example, the set of 'N' individuals are COVID-19 positive at time t and these 'N' individuals COVID-19 positives infected the 'M' other people in the network by the time t + h (i.e., actual spread). To be optimal, the infection spread of a prediction model should be close to the actual spread. Hence, the performance of the models is compared using the infection spread size.

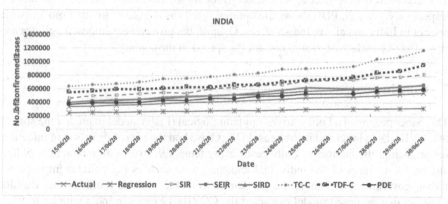

Fig. 4. COVID-19 spread analysis in India15th June to 30th June 2020

As shown in Fig. 4, actual infection spread represented using the blue color line with cross is increasing with respect to time in the actual scenario. The number of confirmed cases in India increases approximately 1.5 times within 15 days. Regression model predicted the spread lower than the actual spread size, and this error in the predicted spread size with respect to actual spread is increasing with time, i.e., True-Negatives. After analysing the infection spread of PDE model, next, experiment depicts the Normalise Mean Squared Error (NMSE) of the prediction size all models for India and all six Indian States.

The Table 5 precisely depicts that the prediction size of the proposed models PDE are better (in terms of Normalize Mean Squared Error) as compared to the other contemporary models, i.e., regression, epidemiological model and diffusion model. It is clear from

Table 5. Approximation error with respect to the actual infection spread size

		Normalise mean squared error (NMSE) (%)						
		Dataset names						
		India	Delhi	Gujarat	Maharashtra	Rajasthan	Uttar Pradesh	Tamil Nadu
Regression		33.89	36.34	12.04	27.64	17.77	57.23	42.35
Epidemiological model	SIR	31.61	24.79	24.17	39.18	34.86	42.45	23.91
	SEIR	17.65	16.99	21.99	23.32	25.02	18.90	17.12
	SIRD	15.76	18.99	18.85	36.46	21.87	18.85	19.12
Diffusion model	TC-C	52.4	48.56	45.23	44.54	41.75	43.91	46.23
	TDF-C	39.4	37.80	35.3	38.01	35.98	37.70	38.51
PDE		9.67	9.66	7.6	8.22	7.54	7.93	8.74

the results that PDE model demonstrated up to 94% accuracy in the prediction of cases in India. Overall, PDE reduced the Normalize Mean Squared Error (NMSE) up to 50%, 35% and 42% with respect to regression, epidemiological model and diffusion model, respectively. Overall, PDE shows the optimal results as compared to other models with respect to India and all six Indian states. Overall, the government-imposed policies are very effective and successful to control the spread in the country.

7   Conclusion

This paper presented a Policy driven epidemiological (PDE) model to predict the spread of COVID-19 in India and its 6 states. PDE is based on the idea of diffusion of infection for predicting the number of daily cases using temporal, structural, interaction, policy and profile features of the individual and the group such as the policies imposed by Indian government to handle the pandemic, available medical facilities etc. Results illustrate that proposed model predicted the COVID-19 cases in India with up to 94% accuracy. The experimental results also reveal that PDE performs considerably better than contemporary model Linear Regression, SIR, SEIR, SIRD, TC-C, and TDF-C models and reduced the Normalize Mean Squared Error (NMSE) up to 50%. In future, proposed model will be used to predict the spread of COVID-19 for the other countries in the world. This work can be extended for the many aspects of COVID-19 impact analysis using heterogeneous features (images and video sharing), such as detection of the source and spread patterns of the virus or medical facility shortage/availability forecasting.

References

1. Gupta, R., Pandey, G., Chaudhary, P., Pal, S.K.: SEIR and Regression Model based COVID-19 outbreak predictions in India. medRxiv (2020)

2. Zhu, H., Wei, L., Niu, P.: The novel coronavirus outbreak in Wuhan, China. Global Health Res. Policy **5**(1), 1–3 (2020)
3. Pueyo, T.: Coronavirus: why you must act now. Politicians, community leaders and business leaders: what should you do and when (2020)
4. https://www.populationpyramid.net/
5. Jiang, F., Deng, L., Zhang, L., Cai, Y., Cheung, C.W., Xia, Z.: Review of the clinical characteristics of coronavirus disease 2019 (COVID-19). J. Gen. Internal Med. **35**, 1–5 (2020)
6. World Health Organization. Coronavirus disease 2019 (COVID19): situation report, 67 (2020)
7. Wu, Z., McGoogan, J.M.: Characteristics of and important lessons from the coronavirus disease 2019 (COVID-19) outbreak in China: summary of a report of 72 314 cases from the Chinese center for disease control and prevention. JAMA **323**(13), 1239–1242 (2020)
8. Nsoesie, E.O., Brownstein, J.S., Ramakrishnan, N., Marathe, M.V.: A systematic review of studies on forecasting the dynamics of influenza outbreaks. Influenza Other Respir. Viruses **8**(3), 309–316 (2014)
9. Nachimuthu, S., Vijayalakshmi, R., Sudha, M., Viswanathan, V.: Coping with diabetes during the COVID–19 lockdown in India: results of an online pilot survey. Diabetes Metab. Syndr. Clin. Res. Rev. **14**(4), 579–582 (2020)
10. Taboe, H.B., Salako, K.V., Tison, J.M., Ngonghala, C.N., Kakaï, R.G.: Predicting COVID-19 spread in the face of control measures in West-Africa. Math. Biosci. **328**, 108431 (2020)
11. Wu, J.T., Leung, K., Leung, G.M.: Nowcasting and forecasting the potential domestic and international spread of the 2019-nCoV outbreak originating in Wuhan, China: a modelling study. The Lancet **395**(10225), 689–697 (2020)
12. Funk, S., Eggo, R.M.: Early dynamics of transmission and control of 2019-nCoV: a mathematical modelling study (2020)
13. Kuniya, T.: Prediction of the epidemic peak of coronavirus disease in Japan, 2020. J. Clin. Med. **9**(3), 789 (2020)
14. Almeshal, A.M., Almazrouee, A.I., Alenizi, M.R., Alhajeri, S.N.: Forecasting the spread of COVID-19 in kuwait using compartmental and logistic regression models. Appl. Sci. **10**(10), 3402 (2020)
15. Mandal, M., Jana, S., Nandi, S.K., Khatua, A., Adak, S., Kar, T.K.: A model based study on the dynamics of COVID-19: prediction and control. Chaos Solitons Fractals **136**, 109889 (2020)
16. Ndairou, F., Area, I., Nieto, J.J., Torres, D.F.: Mathematical modeling of COVID-19 transmission dynamics with a case study of Wuhan. Chaos Solitons Fractals **135**, 109846 (2020)
17. Arora, P., Kumar, H., Panigrahi, B.K.: Prediction and analysis of COVID-19 positive cases using deep learning models: a descriptive case study of India. Chaos Solitons Fractals **139**, 110017 (2020)
18. Wang, L., et al.: Real-time estimation and prediction of mortality caused by COVID-19 with patient information based algorithm. Sci. Total Environ. **727**, 138394 (2020)
19. Gupta, S., Raghuwanshi, G.S., Chanda, A.: Effect of weather on COVID-19 spread in the US: a prediction model for India in 2020. Sci. Total Environ. **728**, 138860 (2020)
20. Agarwal, S., Mehta, S.: Social influence maximization using genetic algorithm with dynamic probabilities. In: 2018 Eleventh International Conference on Contemporary Computing (IC3), pp. 1–6. IEEE (2018)
21. Ceylan, Z.: Estimation of COVID-19 prevalence in Italy, Spain, and France. Sci. Total Environ. **729**, 138817 (2020)
22. Fanelli, D., Piazza, F.: Analysis and forecast of COVID-19 spreading in China, Italy and France. Chaos Solitons Fractals **134**, 109761 (2020)
23. Chimmula, V.K.R., Zhang, L.: Time series forecasting of COVID-19 transmission in Canada using LSTM networks. Chaos Solitons Fractals **135**, 109864 (2020)

24. Cobb, J.S., Seale, M.A.: Examining the effect of social distancing on the compound growth rate of SARS-CoV-2 at the county level (United States) using statistical analyses and a random forest machine learning model. Public Health **185**, 27–29 (2020)

25. Shi, F., et al.: Large-scale screening of covid-19 from community acquired pneumonia using infection size-aware classification. arXiv preprint arXiv:2003.09860 (2020)

26. Tang, Z., et al.: Severity assessment of coronavirus disease 2019 (COVID-19) using quantitative features from chest CT images. arXiv preprint arXiv:2003.11988 (2020)

27. Sarkar, J., Chakrabarti, P.: A Machine Learning Model Reveals Older Age and Delayed Hospitalization as Predictors of Mortality in Patients with COVID-19. medRxiv (2020)

28. Chen, X., et al.: Hypertension and diabetes delay the viral clearance in COVID-19 patients. medRxiv (2020)

29. Sonbhadra, S.K., Agarwal, S., Nagabhushan, P.: Target specific mining of COVID-19 scholarly articles using one-class approach. arXiv preprint arXiv:2004.11706 (2020)

30. Zhang, N., et al.: Severity Detection For the Coronavirus Disease 2019 (COVID-19) Patients Using a Machine Learning Model Based on the Blood and Urine Tests (2020)

31. Hassanien, A.E., Salama, A., Darwsih, A.: Artificial Intelligence Approach to Predict the COVID-19 Patient's Recovery. No. 3223. EasyChair (2020)

32. Barstugan, M., Ozkaya, U., Ozturk, S.: Coronavirus (covid-19) classification using CT images by machine learning methods. arXiv preprint arXiv:2003.09424 (2020)

33. Sethy, P.K., Behera, S.K.: Detection of coronavirus disease (covid-19) based on deep features. Preprints, 2020030300, 2020 (2020)

34. Roda, W.C., Varughese, M.B., Han, D., Li, M.Y.: Why is it difficult to accurately predict the COVID-19 epidemic? Infect. Dis. Model. **5**, 271–281 (2020)

35. Liu, D., et al.: A machine learning methodology for real-time forecasting of the 2019–2020 COVID-19 outbreak using Internet searches, news alerts, and estimates from mechanistic models. arXiv preprint arXiv:2004.04019 (2020)

36. Dandekar, R., Barbastathis, G.: Neural Network aided quarantine control model estimation of COVID spread in Wuhan, China. arXiv preprint arXiv:2003.09403 (2020)

37. Carrillo-Larco, R.M., Castillo-Cara, M.: Using country-level variables to classify countries according to the number of confirmed COVID-19 cases: an unsupervised machine learning approach. Wellcome Open Res. **5**(56), 56 (2020)

38. Magar, R., Yadav, P., Farimani, A.B.: Potential neutralizing antibodies discovered for novel corona virus using machine learning. arXiv preprint arXiv:2003.08447 (2020)

39. Yan, L., et al.: Prediction of survival for severe Covid-19 patients with three clinical features: development of a machine learning-based prognostic model with clinical data in Wuhan. medRxiv (2020)

40. Chen, X., Liu, Z.: Early prediction of mortality risk among severe COVID-19 patients using machine learning. medRxiv (2020)

41. Bailey, N.T.: The mathematical theory of infectious diseases and its applications. Charles Griffin & Company Ltd, 5a Crendon Street, High Wycombe, Bucks HP13 6LE (1975)

42. Peng, S., Zhou, Y., Cao, L., Yu, S., Niu, J., Jia, W.: Influence analysis in social networks: a survey. J. Netw. Comput. Appl. **106**, 17–32 (2018)

43. Zhang, Z., Zhao, W., Yang, J., Paris, C., Nepal, S.: Learning influence probabilities and modelling influence diffusion in Twitter. In: Companion Proceedings of the 2019 World Wide Web Conference, pp. 1087–1094 (2019)

44. Agarwal, S., Mehta, S.: Effective influence estimation in twitter using temporal, profile, structural and interaction characteristics. Inf. Process. Manag. **57**(6), 102321 (2020)

45. World Health Organization, & World Health Organization. Report of the WHO-China joint mission on coronavirus disease 2019 (COVID-19) (2020)
46. Acuña-Zegarra, M.A., Santana-Cibrian, M., Velasco-Hernandez, J.X.: Modeling behavioral change and COVID-19 containment in Mexico: a trade-off between lockdown and compliance. Math. Biosci. **325**, 108370 (2020)

Author Index

Printed in the United States
by Baker & Taylor Publisher Services

Printed in the United States
by Baker & Taylor Publisher Services